VIEW FROM THE COCKPIT

C000265348

TIM MCLELLAND

VIEW FROM THE COCKPIT
FLYING MILITARY AIRCRAFT

The
History
Press

Front cover illustrations, from top: Luftwaffe Eurofighter Typhoons.
(Eurofighter); Royal Air Force Typhoon pilot. (Tim McLelland)
Back cover illustration: A gaggle of Jaguars make their return flight to
RAF Coningsby. (Tim McLelland)

First published 2014

The History Press
The Mill, Brimscombe Port
Stroud, Gloucestershire, GL5 2QG
www.thehistorypress.co.uk

© Tim McLelland, 2014

The right of Tim McLelland to be identified as the Author
of this work has been asserted in accordance with the
Copyright, Designs and Patents Act 1988.

All rights reserved. No part of this book may be reprinted
or reproduced or utilised in any form or by any electronic,
mechanical or other means, now known or hereafter invented,
including photocopying and recording, or in any information
storage or retrieval system, without the permission in writing
from the Publishers.

British Library Cataloguing in Publication Data.
A catalogue record for this book is available from the British Library.

ISBN 978 0 7524 9002 1

Typesetting and origination by The History Press
Printed in India

CONTENTS

INTRODUCTION

It's probably fair to say that as a devoted aeroplane enthusiast, I've been remarkably lucky. Like so many others, I grew up with a passion for aviation, and spent as much time as I could lurking around the perimeters of airfields or making my way to whatever air shows I could get to. When I started working as a photographer and a writer, I had the chance to take my enthusiasm further and visit some of the bases and units that were the subject of my fascination. Not only could I get a chance to talk to the people who fly and support the warplanes that I worshipped, I could even go and fly with them, and see for myself what military flying was all about. Of course, times have changed, and a lack of resources and money means that military air arms now only rarely invite civilians to climb into their aircraft and go flying. Likewise, the many and varied warplanes of the Cold War era are mostly gone and there isn't much that is left to capture the attention and imagination of an enthusiast. So by any standards I was lucky to have found myself flying in so many aircraft that I'd studied and obsessed about for so long.

In this book, I've tried to share some of my enthusiasm for the benefit of those who would, I'm sure, have loved to have enjoyed the same experiences that I had. I don't have the ability to take everyone flying but I can do my best to describe at least a little of what military flying is like. Hopefully, the following chapters will give a taste of what it's all about, and allow the more imaginative reader to mentally put himself in the cockpit (or the flight deck) of a few magnificent machines. It would have been possible to describe many other aircraft in addition to those that have been included, but there is always a risk of repetition, and the limits of space mean that it is probably better to describe a few aircraft in some detail rather than a lot of aircraft in more general terms. I hope the following pages convey at least some of the fascination and excitement that I attach to all aspects of military aviation, and the same 'buzz' that I share with all of my fellow enthusiasts out there.

I should also take this opportunity to thank the countless people who have, over many years, enabled me to make all of my visits, all of my flights, conduct all of my interviews and much more. Their patience and helpfulness are of course greatly appreciated. I must also extend my thanks to fellow 'plane freak' Jeff Eddy for his help in creating some of the material that follows.

Tim McLelland

RIGHT: The author strapping in at the beginning of a low-level Hawk sortie from RAF Valley in Wales. (*Glenn Ashley*)

1

LEARNING THE BASICS

A good place to start with a look at military flying is with the basics. RAF Swinderby was once a busy and very significant bomber base during the dark days of the Second World War. As a former home to Lancasters, the station remained active through post-war years but slowly declined in importance until flying activity ended completely. Aeroplanes returned many years later, however, when the RAF selected Swinderby as the home to a very active unit that was responsible for the training of newly qualified RAF officers who were embarking on careers as potential pilots. The once busy airfield was by then largely deserted but the enclave of brightly painted De Havilland Chipmunk trainers soon became a familiar site in the skies south of Lincoln. The EFTS (Elementary Flying Training School) was tasked with the responsibility of introducing the RAF's newest pilot recruits to the world of flying, as described by one of the unit's instructors:

A small number of each new officer intake would have had sufficient flying experience to bypass the EFTS altogether, and they were sent directly to a BFTS [Basic Flying Training School] to begin flying the Jet Provost. To be eligible for this shortcut, one must already have had at least 30 hours of flying experience. A significant proportion of students who met this criterion came directly from the Royal Air Force's UASs [University Air Squadrons]. There were [during the 1980s] sixteen UASs located around the UK, responsible for the

flying training of university students who had expressed an interest in joining the RAF. There were two ways in which young men [and women] could apply to be accepted into a UAS. Firstly, there was a university cadetship scheme, whereby the student committed himself to an RAF career, and was thus commissioned as an acting pilot officer [APO], after undergoing selection at Biggin Hill and a short preliminary training course. The student was paid a salary and, upon graduation from university, moved directly into the RAF. Other students joined the UAS as cadet pilots [CPs], after passing interviews, and although the UAS may have sought to accept those with an interest in the RAF and aviation, CPs did not have to commit themselves to an RAF career. A CP could expect to spend up to two years with his UAS, on a strictly part-time basis, as the RAF recognised that the UAS activities had to take second place to the university academic studies.

The amount of flying time on what was then the UAS basic trainer – the Scottish Aviation Bulldog – generally reached about sixty-five hours after two years, while APOs generally completed a third year with the UAS, finally leaving with at least ninety-five hours of experience on the Bulldog. Roughly 30 per cent of any squadron's membership would be cadet pilots, and from these, perhaps half a dozen would ultimately decide to join the RAF, while the majority of the remaining personnel would be APOs, already committed to their career.

LEFT: Chipmunk WB551 is not all that it seems. This aircraft was a Canadian-built example operated by De Havilland prior to the completion of the first example for the RAF, WB549, which made its first flight shortly after this photograph was taken in 1949. (*Tim McLelland collection*)

The new arrivals at Swinderby often found themselves renewing an old friendship with the faithful 'Chippie'. Many aspiring pilots had inevitably joined the Air Training Corps (ATC) during their school days, and the RAF operated a large fleet of Chipmunks on behalf of the ATC. Almost every ATC cadet was given an opportunity to fly in a Chipmunk at a local airfield, and thus this venerable trainer was no stranger to many students. However, flying the Chipmunk at Swinderby wasn't simply a source of enjoyment as it was in the students' ATC days. The function of the EFTS was to give the students a basic instruction in the techniques of flying an aeroplane, effectively serving to weed out those who were physically and mentally unable to come to terms with the task. Until 1987, students were sent to Swinderby on a six-week course, consisting of fifteen flying hours on the Chipmunk, simply to be assessed as potential pilots. The FSS (Flying Selection Squadron) operation worked well, but further studies indicated that this first introduction into the flying environment could be used even more constructively, and so the EFTS was expanded to actually teach, rather than just select. The EFTS instructor continues:

Obviously the RAF is keen to establish at an early stage which students are likely to make successful pilots, as the flying training process is a long and expensive undertaking, and it is far better to withdraw those who are identified as training risks sooner rather than later. Estimates in the 1980s suggested that the total cost of training a fast-jet pilot is approximately £3 million and, as one might expect, great care is taken to ensure that such large sums of money are spent very wisely. Thus, the role of the EFTS is quite an extensive one, aiming to give each student a thorough training on an inexpensive aircraft, during 64 hours of flying spread over sixteen weeks. Normally eight or nine students form the basis of each of the new courses, which commence on six-weekly cycles. We test the ability to absorb instruction more than anything else, so that we can teach the student some skills, but also assess his ability to learn. We don't attempt to grade the students, but if they don't come up to standard they are still chopped from the course, certainly. They still have to pass each stage.

The EFTS instructors were always keen to explain the advantages of their sixteen-week course:

What we effectively have at Swinderby is a low-cost flying training school that acts as a preliminary stage before the main flying training phase. Before the new system started, the students went away and moved straight on to the Jet Provost, which was an expensive way to do things. In effect we indirectly grade students here, if you like, and it's fair to say that if you can get through EFTS you're likely to be successful on the Jet Provost. The new system certainly is cost-effective. We understand that the saving over the old methods is about seven hundred thousand pounds per year, which is quite significant, so the EFTS makes sense both practically and financially.

The student's first flight came after a period of ground instruction, and was essentially a familiarisation flight lasting roughly forty-five minutes. On the second flight, the Effects of Controls teaching began, demonstrating the basic use of rudder, ailerons, elevators and throttle. The student, who sat in the front seat, watched as the instructor demonstrated the effects of the control surfaces and throttle, before repeating each exercise as directed by the pilot. The introduction was a patient and gentle one. By the fourth flight the student was being introduced to other activities, such as 'Straight and Level' (keeping the aircraft on a steady course, using the controls and trim), and taxiing the aircraft on the ground. The art of taxiing along an airfield runway or perimeter track was by no means simple in an aircraft such as the Chipmunk. With its tailwheel undercarriage, the pilot did not have a direct forward view and had to rely on a Spitfire-style 'weave' along the taxiway, visually checking port and starboard clearances in turn, in an attempt to negotiate the airfield successfully without hitting anything. The gentle use of brakes and power was a skill that had to be carefully learned, although the expanses of the relatively uncluttered (largely deserted) and traffic-free airfield at Swinderby certainly made life easier for the students.

By Exercise Seven, the variety of aerial activities had increased further, to include climbing and descending turns and stalling:

> We gradually expand the exercises, handing more and more responsibilities to the student, so he gets more and more to do. The instructor is there to resolve the mistakes, but the basic job is to demonstrate and then let the student repeat the action. He sees the instructor do it first, then he has a go himself.

RIGHT TOP: A trio of Chipmunks from RAF Swinderby's EFTS, pictured just a few miles from home base, overflying the city of Lincoln. (*Tim McLelland collection*)

RIGHT BOTTOM: Despite having officially retired from RAF service, two Chipmunks are still flying with the RAF. WK518 operates with the Battle of Britain Memorial Flight as a communications aircraft and trainer. It now wears a dazzling silver and fluorescent orange paint scheme, replicating the colours worn by RAF Chipmunks in the 1960s. (*David Whitworth*)

The basic skills were learned quickly, but often after a great deal of work:

> Things like accelerating and decelerating aren't as straightforward as they sound. It's like a child learning to walk for the first time, just simple things like learning to use your feet on the rudder in association with the throttle, keeping straight and level.

Each flight included revision of techniques learned in previous exercises, and thus the learning curve pointed remorselessly upwards. Take-off and landing practice and flying around the airfield circuit became exercises that were repeated time and time again, in preparation for the first main hurdle the student had to face – his first solo flight.

The first solo generally took place in Exercise Thirteen (the EFTS instructors were apparently not superstitious), and the student was simply required to fly the Chipmunk safely around the airfield circuit, unaided by the normal assurance of a voice from the aircraft's rear cockpit. It sounds like a simple task, and in many ways it was, but for the student pilot it was a hugely important milestone:

> Nobody thinks they will ever go solo, but they inevitably get there. What we have to ask ourselves as instructors, is whether the student is capable of going solo, whether he can simply cope on his own, without getting into difficulties or doing anything dangerous. The Chipmunk isn't the most difficult aeroplane to fly, but it is quite demanding. It needs to be worked at all the time and it's not an aeroplane that you can just sit back in and let it do its own thing.

For anyone who was unaccustomed to flying, the first ride in the Chipmunk was a great thrill, and the author was invited to sample a flight to see for himself. After settling into the tiny cockpit, complete with parachute pack, life jacket and flying helmet, the tiny Chipmunk seems even smaller, and although the aircraft was (and is) a sturdy old thoroughbred, the airframe creaks, rattles and rocks on its undercarriage legs, seemingly too light and delicate to fly. But when the engine starts, the little 'Chippie' begins to sound like it means business. The 145hp piston engine is started by cartridge, the crack of the starter and the smell of the cordite smoke being familiar to countless RAF pilots who

learned their skills on this, the most famous of the RAF's basic trainer types. Progressing to the runway was a skill that students worked hard to learn but for the very first flight it was always the instructor who demonstrated that once the art of tailwheel manoeuvring is mastered, the Chipmunk could be controlled quite easily. Chugging along the old bomber taxiway, the tail swings to the left and right so that my pilot can see forward and guide the machine out to the runway. Out on the expanse of one of Swinderby's long bomber runways, the diminutive 'Chippie' is afforded a very generous amount of space in which to get airborne. Normally accustomed to grass fields, the Chipmunk needs very little distance or effort to get into the air, the tailwheel quickly coming up as soon as the brakes are released and the engine is brought up to full power. In just a few seconds the aircraft gathers enough speed to fly, usually becoming airborne at about 60 knots, before gently climbing away at 70 knots into the fluffy clouds, seemingly slow and laborious.

ABOVE: WB550 is high above the clouds during a photo shoot in the 1950s. This aircraft eventually joined the EFTS at Swinderby, prior to leaving RAF service and becoming a civilian-owned example. (*Tim McLelland collection*)

Nothing ever happened too quickly in the Chipmunk, and this was one of the aircraft's assets. To demonstrate a typical circuit, at about 800ft, the aircraft is gently turned into the downwind leg (largely with rudder input) and round on to finals, with flap selected down, and speed dropping away to just 50 knots as the Chipmunk drifts back over the runway threshold, the engine idling. Some careful control was always needed to keep the aircraft straight on the runway, and particularly on take-off there was always a distinct tendency to swing off the runway in response to the torque effect of the propeller – a characteristic common to piston engine tailwheel aircraft. However, in most respects, the 'Chippie' was the perfect vice-free basic trainer, requiring plenty of attention but only the most basic of flying skills. If any proof were needed that the Chipmunk was indeed an ideal primary trainer, one only has to note that the aircraft was in service with the RAF from 1949, and continues to fly with the RAF Battle of Britain Memorial Flight as trainers for the much bigger and powerful machines that the unit operates, all of which employ the same piston engine power and tailwheel configuration. One of the EFTS Chipmunks had the distinction of being the oldest aircraft in operational service. WB550 entered service in November 1949 and could be seen flying at Swinderby from 1974, finally leaving RAF service in 1986, although it remains active in civilian hands even to this day.

The EFTS instructors were experienced members of the RAF, and could boast many years of service operating a wide variety of aircraft types around the world. The QFIs (Qualified Flying Instructors) generally worked with specific students throughout the duration of their courses, and the EFTS didn't see much value in pooling their resources:

> We like to put one instructor with one student to assist in the learning process. It helps if the student has to relate to just the one instructor. It also helps to give a better judgment of how a student performs overall. If the instructor kept changing, it would be difficult to get a full picture of how good or bad a guy might be. If, however, a student suddenly finds that life is becoming difficult, the staff will change the QFI, in order to be sure that the student isn't suffering

from a personality clash with the instructor. We will change to a different instructor, and we are able to give a few more hours to sort out any problems, if we think this is necessary.

Giving a student extra attention was referred to as being 'on review'. The QFIs regarded this process simply as an intensified period of instruction, and it didn't necessarily mean that the student was going to fail. The student, however, inevitably viewed such action with great suspicion, and often feared that their stay with EFTS wasn't going to result in success. There was a certain amount of mental pressure built into the course, chiefly because the flying exercises were constructed to expand the abilities learned in the preceding sortie:

> You find yourself learning one skill, and the next time you fly, you're expected to do that without any problem, and to be able to cope with something new. When you start, you think, bloody hell, I'm never going to make it, but I'm sure everyone enjoys the course … we live on adrenalin here.

For about 25 per cent of each intake, though, the EFTS proved to be an insurmountable hurdle and their military flying career could suddenly be over:

> For some guys it's a big blow, a cherished dream since childhood gone out the door … We explain everything of course, but even so, you can't take on a guy who's going to go out and kill himself.

Halfway through the course, the capabilities of each student pilot increased enormously. Having come to Swinderby with no flying experience, they would now be learning to put the Chipmunk into spins, steep turns, barrel rolls and stalls. They would have started to move away from the airspace in the immediate vicinity of Swinderby, navigating as well as flying. The basic skills of control would now be taken for granted, and each flight involved more complex manoeuvres, together with additional responsibilities, such as fuel monitoring, radio communication, and emergencies, such as an 'EFATO' (engine failure after take-off):

We will expect the guys to put the content of the sortie together in an intelligent manner, not just fly the aircraft. Things never happen the way you plan them in flying, and the pilot has to be able to look after himself. However, we don't expect anyone to be perfect ... we learn where the imperfections are. Some come here never having flown before at all, and it can be harder for them to settle down, but all we ask is that they make reasonable progress. The failure rate isn't high at all. If there is one thing that causes a student to fail, it's the lack of ability to cope with more than one thing at a time. They might not be able to change speed and direction, and talk on the radio at the same time. Some just can't hack it. It's not a continual test though, it's more a matter of developing abilities ... We're here to help the guys, not chop them, and we do our very best to get the ability out of them.

The course was divided into sections, each containing a number of specific skills which were then taught to the student. The skills are practised and finally tested at specific points ('critical points') or during the Flight Commander's Check Flight and the Final Handling Test at the end of the course. The few failures that did occur, tended to happen at the first solo stage, or during the Flight Commander's Check. The length of each sortie was laid down in the course syllabus, generally averaging about one hour. Students normally flew three sorties per day, although much depended on weather conditions, and sometimes a whole day could be lost to bad weather. The QFIs worked even harder than the students, and could total four sorties on some days. Even when conditions prevented flying, the learning process continued on the ground, in lectures or private study:

There are exams to pass, and the ground instruction is as important as the flying. The meteorological package includes education on all aspects of weather, such as the atmosphere, clouds, winds and so on. We have to explain what the sky can do to you ... It's very important when you could literally have your wings torn off. You have to avoid thunderclouds, hail, dangerous air currents and so on. Really it's a mini weather forecaster's course ... a lot of very dangerous conditions exist ... for example things like airframe icing can be lethal.

The second half of the course continued to develop the basic skills, while introducing more complications. Navigational exercises became more elaborate, with triangular routes, and solo flights away from the airfield. Instrumental flying came into the syllabus, with instruction on how to use the complete instrument layout and 'limited panel' flight. The basics of formation flying required the students to develop the skills of station-keeping, breaking and rejoining formation, and turning in both line-astern and echelon formation. Eventually the student would fly formation solo, keeping station with the instructor's aircraft, through a series of climbing, descending and turning manoeuvres. Navigational skills culminated in unplanned diversions, requiring instant decisions from the pilot:

The Chipmunk can be tricky to fly accurately, which is why it makes such a good trainer. It develops the necessary skills ... it's quite a testing little aeroplane ... it's been used to train for the Spitfire, and it does the job at EFTS very well. Look at WB550 which has been around since 1949, it was the second production aircraft, so it's really an ancient piece of kit, but as good as ever. We wouldn't take something like a Cessna, simply because it would make life too easy ... you'd stooge around like a car driver, which is no good at all. The Chippie has no great merit, but no vices either. The Bulldog trainer could chop through cloud and do something on top of it certainly, whereas we're limited in that respect ... if the cloud is low, we can't fly. We do teach a great deal to the students, but it's only enough to get a foothold on the ladder though ... airmanship is important, just learning the rules of the air. Instrument flying involves explaining and demonstrating how all the dials work. They have to learn to navigate and map-read on cross-country flights, and you expect them to be able to fly to a destination and know the arrival time, fuel state and so on. They must plan a course and fly it, even with things like wind changes ... some just sit there dumbstruck. We look at each skill and try to determine if they have 'em all, basically!

The final flight in the course was the FHT – the Final Handling Test. Rather as in a driving test, the examiner would require the student to perform a series of manoeuvres during a one-hour sortie:

The FHT really encompasses everything in the course, and the student will have to fly the aircraft well and perform the basic aerobatics and navigation exercises we've taught him, as well as cope with emergencies and so on. For the aerobatics, he'll be expected to do loops, barrel rolls, slow rolls, rolls off the top of a loop and so on … he'll be expected to put them all into a sequence in an intelligent manner. He'll have to look after the navigating himself and comply with the usual flying rules, not flying below certain heights, etc. We'll throw in things like radio failures, closing the throttle … all sorts of things. There's no set rule except that there will be surprises … it keeps you sharp and on your toes … various emergencies will be thrown at him, like oil pressure falling, engine fire, all the things which need a decision.

Flying and academic exercises aside, the students also had to come to terms with their new responsibilities as officers:

It's not as bad here as at Cranwell, where you're watched all the time, but we're still judged on our officer qualities … things like looking after the desk in here, and the Battle of Britain day, when we're the hosts … You aren't under any great pressure, but you have the moral obligation not to let the side down … It's much more relaxed here, though.

The successful students left Swinderby to join one of the Basic Flying Training Schools, putting their newly acquired skills to work on the Jet Provost and, although the prospect of flying a jet aircraft might have been rather daunting to the EFTS students, the instructors were confident that the training at Swinderby ensured that they would be able to cope with the transition:

The Jet Provost training format is essentially the same, simply an expanded version, but with the same basics. The QFIs here at EFTS are all very experienced, and we've all instructed on other aircraft, so we know how the training system works. We all have experience in different areas.

Sadly, the Chipmunk's association with the RAF is now almost at an end, and the dwindling amount of elementary flying training that is still conducted is flown in aircraft of a far more recent era. RAF Swinderby too has gone, and the sound of the little Chipmunks is consigned to history, much like the roar of Swinderby's Lancasters that preceded them.

The fledgling flyers who survived Swinderby's demanding training regime moved on to fly the Jet Provost, a simple, straight-winged jet designed as a direct development of its piston-powered predecessor. My personal affection for the Jet Provost goes back to my childhood, when I first encountered the aircraft at the annual Battle of Britain 'At Home' day held at RAF Finningley (now Doncaster Sheffield Airport) every September. Every show through the 1960s and 1970s included a demonstration of solo Jet Provost aerobatics and often also included a display of formation aerobatics, courtesy of the various Jet Provost display teams that appeared, such as the 'Red Pelicans', 'Poachers', 'Blades', 'Gemini Pair' and others. I have to admit that my interest in the Jet Provost was shamelessly limited back in my childhood, as my attention was inevitably drawn towards the speed and agility of aircraft such as the Lightning, or the ear-splitting noise of the Vulcan bombers. The lowly Jet Provost was, by comparison, not much of a thrill when it didn't exhibit any of the speed or noise of its operational counterparts, but time certainly changes perceptions, and in retrospect I can't help wishing I could see those magnificent formation displays all over again. It was in the early 1970s that I got to see yet more Jet Provosts, courtesy of the former wartime airfield at Strubby in Lincolnshire, where the RAF's College of Air Warfare regularly exported much of its aircraft traffic from home base at nearby RAF Manby. For a young plane spotter, the black and yellow wooden perimeter fence that stretched around Strubby's runway was a fascinating place to be, and amongst the daily appearances of Varsities and Dominies, the little Jet Provosts also joined the airfield circuit. If one was very lucky, one could even catch a day when the CAW Jet Provost display team (the 'Macaws') showed up to practise their manoeuvres above Strubby. Nothing could be more captivating than the sight of a gaggle of Jet Provosts looping and rolling through a sunny Lincolnshire sky.

ABOVE: The first of the many; XD674 was the very first Jet Provost T1, making its maiden flight on 26 June 1954, marking the very beginning of the Jet Provost's long and distinguished service with the RAF. (*Tim McLelland collection*)

It was many years later however when, almost by chance, I found myself getting airborne in a Jet Provost. I was commissioned to photograph the RAF's 'Vintage Pair' Meteor and Vampire, and we needed a suitable aircraft from which to take aerial photographs. The Jet Provost was an obvious choice, as the operators of the Vintage Pair (the Central Flying School) had a sizeable fleet of Jet Provosts at that time. My experience of flying was then limited, to say the least, and the prospect of taking to the skies in a Jet Provost was quite a thrill. My excitement was tempered slightly when I actually found myself on RAF Scampton's runway, strapped to the starboard seat of a Jet Provost T5, with a Meteor and Vampire roaring skywards ahead of me. It was only then that I realised that three RAF jets were about to fly just for me, and I would have to get some good photographs in exchange for such an expensive and generous act; therefore it seemed inappropriate to be dwelling on the sheer fun that appeared to be in prospect.

ABOVE: Jet Provost XM384 survived only a few more weeks after this publicity photo was taken. On 26 May 1966, it collided with another Jet Provost during formation flying whilst serving with No 2 FTS. The two crew members ejected safely from the aircraft. (*Tim McLelland collection*)

But once airborne and high in the sky over Lincolnshire, it was impossible to avoid the feeling of great joy as the little Jet Provost demonstrated its docility and reliability, while my pilot carefully manoeuvred us around our two targets and I grabbed my photographs. By the end of the sortie we'd rolled, looped, climbed and descended, in a manner that I'd never experienced before, but none of this seemed particularly terrifying inside the trusty Jet Provost; in fact it was undoubtedly great fun. The photographs were good too.

I renewed my acquaintance with the Jet Provost (this time in an earlier variant of the same aircraft, the unpressurised T3A) at Church Fenton, from where No 7 FTS operated another substantial fleet of aircraft. One of the unit's QFIs treated me to a low-level (250ft) flight out to the coast off Scarborough, at which stage we indulged in some aerobatics high over the sea (all of which is described in this chapter). Other assignments with the Jet Provost followed, the very last being surprisingly similar to the first, but this time with the magnificent Vulcan XH558 as my quarry. Once again I was airborne in a Jet Provost Mk 5 from Scampton, this time heading south at low level on a warm, calm summer's morning, scooting over the rooftops of Lincoln to nearby Waddington, where we briefed our photo flight with the Vulcan. For the actual photo flight, one of the most memorable highlights occurred before we even left the ground; lined up on the runway, XH558 was positioned directly behind us and by glancing up at the cockpit rear-view mirror, I could see the menacing grey shape of the huge Vulcan, seemingly bearing down upon us from behind. I never established what the aircraft photographers thought of us as they stood in their familiar spots on the nearby A15 road, presumably wondering why a huge Vulcan was following a tiny Jet Provost along Waddington's long runway.

Flying formation on the Vulcan was a challenge for both the Jet Provost and its pilot, as the acceleration and deceleration of the mighty Vulcan is significantly different to that of the diminutive Jet Provost, and maintaining a decent position for photography wasn't easy, especially when getting in too close was far too risky to contemplate; my pilot reminded me that the ill-fated XB-70 Valkyrie had been destroyed thanks to a chase aircraft getting a little bit too close and being pulled down onto the aircraft's huge wing surface … we didn't want to repeat history with the Vulcan. My experiences with the Jet Provost were all good ones,

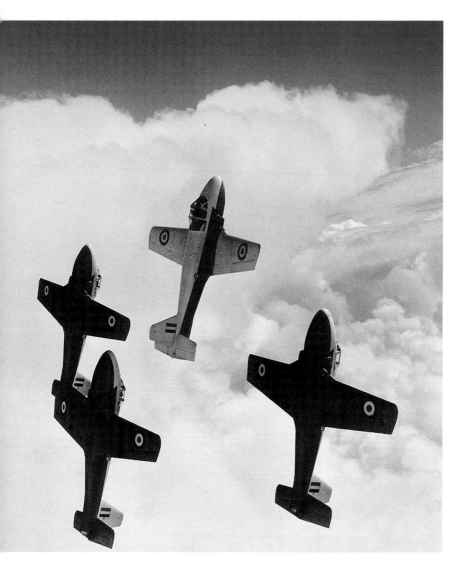

LEFT: The Jet Provost entered RAF service in 1957, although acceptance trials began in 1955. The Central Flying School quickly formed an aerobatic display team with four Jet Provost T1 aircraft in 1957, operating as 'The Sparrows' through 1958. (*Tim McLelland collection*)

The long lines of red-and-white-painted Jet Provosts at the RAF's Flying Training Schools were a familiar sight for decades, and the skies above Linton-on-Ouse, Church Fenton, Leeming, Cranwell, Syerston and Little Rissington often seemed to be alive with Jet Provosts from dawn until dusk … and even longer sometimes. It was at these bases that the RAF's potential pilots got their first chance to take the controls of a jet aircraft, and the prospect of getting airborne in the Jet Provost inevitably excited every new student as he arrived at his new posting. Unfortunately for them, the first five weeks at a Basic Flying Training School were spent firmly on the ground in a period of concentrated education concerning the technical aspects of the Jet Provost, as well as a great deal of time studying more general issues of airmanship. The facts and figures which needed to be memorised were many and varied: engine start-up and shutdown, engine flame-out, engine re-lights, hydraulic malfunctions, bird strikes, radio failure, electrical systems, cockpit instrument layout, airframe icing and much more. Before a student could take any aircraft into the air, he obviously had to understand the ways in which the aircraft functioned and how it must be operated. The RAF's QFIs were always happy to explain how the training system operated:

The Ground School runs on a sort of package basis, in that it is the same for every course, and is designed to fit into an initial period of five weeks. The students work at their own pace, which includes a great deal of evening study outside of the classroom, as there's a lot to learn. In fact there's about two hundred hours' worth of work to take in. After completion of this intensive academic period, the course concentrates on flying instruction, the initial period being aimed at getting the student through the first major hurdle on the course – his first solo flight in the Jet Provost. The basics of flying are re-applied, exploring Effects of Controls, take-off and landing. Gone is the piston engine that they will be familiar with from their initial selection and training, and gone too is the Chipmunk's tail-wheel. For the first time,

and like all of the countless crews that learned to fly in the Jet Provost, I regard the aircraft with a particular affection and nostalgia. It was an unremarkable machine, but that was the key to its success. It was simple, easy to fly, surprisingly fast and nimble, and reliable. The RAF disposed of its Jet Provosts not because of any deficiencies but simply because they were so old that new aircraft had to be brought in to replace them.

the student is sitting on an ejection seat, equipped to punch him out of the cockpit upon initiation. So, having arrived at the FTS, they'll first go to the Ground School, but during this period they will be given a familiarisation flight, and we'll take them into the air for half an hour or so, just so they can see what it's going to be like.

The instructors have a typical matter-of-fact attitude, but for the students, their first flight in a Jet Provost is a thrilling experience, as it will inevitably be the first time that they've flown in an aircraft that possesses speed, agility and a jet engine.

The Jet Provost isn't a particularly big aeroplane, nor is it one that captures the imagination. By any standards it's a simple design, functional and hardly indicative of the world of high-speed flight that the students are hoping to join. But once inside the Jet Provost's relatively roomy cockpit, tightly strapped to an ejection seat, with the rumble of a jet engine just a few feet behind one's head, there's more than a little anticipation when the first flight is about to begin and the QFI's voice crackles through the RT: 'Okay, seat pins are out and stowed, and we're ready to line up on the runway. If we do have to eject, always go for the seat pan handle, as it's very difficult to reach the face-blind handle on the top, behind your head.' When just a second could mean the difference between surviving or being killed, it does matter. 'On the command "Eject, eject, eject" get your head well back into the seat headrest, and give that handle a good hard pull.'

For the student, looking out at the world through a visor, over an oxygen mask, from inside a heavy and cumbersome helmet, the conditions are very different from those in the diminutive Chipmunk and Bulldog. The actual take-off isn't very impressive however, as the Jet Provost doesn't boast a particularly powerful engine, and the acceleration is gentle and almost laborious, accompanied by a crackle of engine noise that slowly merges into a gentle whistle and a rush of airflow as the aircraft gradually picks up speed.

'Okay, we're rolling now … it's accelerating, I assure you … A little pressure on the stick and we're airborne. Undercarriage coming up and we're nicely settled in a gentle climb.' Safely into the air, the little Jet Provost settles into a typical transit speed of 250 knots (nearly 300mph) and an almost standard low-level altitude of 250ft. The RAF embraced low flying almost from its very beginning, but from the 1960s onwards it became the only means of ensuring survivability in a hostile environment, hidden from the eyes of enemy radar. For the earliest stages of aircrew training, students were introduced to the lowlevel environment and the kinds of altitudes that they would adopt on most of their operational missions, should they become combat pilots. Looking out from the Jet Provost's bubble canopy, one can see the world flashing by – fields, villages, rivers, hills and much more, in a seemingly endless flow. It's exciting but certainly not frightening as the Jet Provost provides enough performance to give a good feel for what high-speed flight is like, but not so much as to overload the senses of a novice pilot. Aerobatics offer an even greater thrill and after climbing to altitude, the instructor turns the little Jet Provost through a wide and lazy circle, checking the surrounding airspace for conflicting air traffic or areas of bad weather. The sky is a crisp, deep blue and completely clear of all potential hazards. It's a good place to stay for a while to admire the panoramic view, but suddenly the world turns upside down.

ABOVE: This manufacturer's publicity photo illustrates the simple sliding canopy employed on the early versions of the Jet Provost. Rubber grip patches can be seen stuck to the upper surface of the air intake, designed to enable crews to step safely into the cockpit. (*Tim McLelland collection*)

'Into a slow roll now … over we go … here comes the horizon … round she comes … Okay? … don't forget to let me know if you're feeling unhappy.' The aircraft's nose gently pulls upwards and the landscape slips away under the aircraft and behind the wings, as the Jet Provost settles into a steady vertical climb, pointing directly upwards into the deep blue. Glancing inside the cockpit at the black instrument panel, the airspeed indicator confirms that the forward speed is decreasing rapidly, and while the instructor kicks the rudder foot pedals, the nose begins to gently swing left and right.

'Now, which way is she wanting to go? … Okay, we're nearly out of airspeed now and we're turning to starboard … and over goes the nose.' For a brief moment, everything is stationary. The forward motion of the Jet Provost is perfectly matched by the pull of gravity, and the student and instructor hang motionless in mid-air, until the rudder inputs push the aircraft over into a stall turn, flicking it around into a vertical dive, pointing directly down towards the distant countryside below. But things get even more disorientating.

'Just a little pressure on the stick, and round we go,' says the QFI, and the forward view of the ground begins to rise as the aircraft's nose edges even further downwards (but which way is down?) and the Jet Provost is pushed into an inverted oblique descent, the ejection seat straps and small dirt particles suddenly straining to push through the canopy instead of the cockpit floor. The sensation of negative-g is peculiar, if not unpleasant. Most people's experience of the gentle application of negative-g when driving over a hump-backed bridge is okay, but this is very different. It's far more severe and doesn't abate after a couple of seconds. Instinct tells you to hold on to the cockpit wall, just in case you're somehow pulled out of your seat, upwards (or should that be downwards?) through the Plexiglas canopy. Of course the straps keep you firmly in place. But tell that to your senses …

A gentle roll brings the Jet Provost back into a more comfortable erect descent, the fields below (now thankfully back in their conventional position) begin to look pretty close, and as the bottom of the dive is reached, the Jet Provost is pulled into level flight, the gforce (now a positive sensation that pushes the pilot hard into the aircraft's seat) begins to increase. With a pretty heavy flying helmet (the 'bone dome') to support, it's an effort to keep your head upright. Students soon learn not to get caught with their head down when gforce is applied, as they would be unlikely to get heads-up again until the manoeuvre stops. The accelerometer reads '3' … only a mere 3g and for a new student, that might seem impressive, but it's nothing, and it's only a taste of the brutal applications of five, six or seven times the normal force of gravity that can regularly be applied during combat flying.

'The clock says we should be heading home now … how about you take the controls now? … you have control … that's the heading we want, so try and keep us straight and level.' In the Jet Provost, maintaining a steady path is not much of a challenge, and although the controls are fairly relaxed (QFIs often made wry comments about moving the control column and then waiting for something to happen), the Jet Provost responds well to inputs. It is a docile machine with no significant handling issues, and for a student it was an ideal aircraft in which to learn the skills of basic flying. Some suggested that it was perhaps too docile, and its forgiving handling characteristics were arguably unrepresentative of the kind of challenges that students ought to have been exposed to. But by any standards the faithful Jet Provost was a classic basic trainer that could take a great deal of punishment or mishandling without danger.

With the sortie completed, the Jet Provost is soon back over the airfield, and a relatively fast arrival is followed by a tight turn into the airfield circuit, the airspeed rapidly bleeding away as the aircraft rattles around onto a steady downwind leg. With the speed reduced, down come the landing gear and flaps, and the aircraft is gently brought around onto final approach. With flaps down and gear extended, the little Jet Provost gently rocks and bumps its way back towards the runway, and for the student this is the first taste of countless practice landings that will follow, as each fledgling flyer learns to position the aircraft on approach with the right speed and right height, gently easing the throttle back and forth to adjust the aircraft's height whilst making small moves on the control column to raise or lower the nose in order to maintain the right landing speed. In a matter of seconds the runway threshold suddenly begins to rush towards the aircraft and with a gentle ease back on the control column, the Jet Provost's main wheels ease back on to the concrete. The delights of these introductory flights serve to fill the students with enthusiasm for the long and demanding training period that lies ahead:

On the actual course, we start off with the basic teaching exercises like Effects of Controls, moving on to straight and level, just learning how to maintain straight and level flight. The teaching is slowly built up and a full briefing before each flight is always necessary. The students will go off into a cubicle and they'll cover all the aspects of the flight, the type of exercise, and basic airmanship points … it's rather like if you're in a car, just establishing basic road-sense. It's not just the mechanics of flying, it's being aware of all the other factors, such as lookout, who is in control of the aircraft, orientation and system checks like oxygen, fuel, and that sort of thing.

The students are tasked not only with the control of the aircraft but also the management of the on-board systems too, chiefly the fuel supply, but many other things as well. As the exercises progress, more and more responsibility is put directly on to the student:

On the initial sorties they will be told to check fuel about every five or ten minutes, but the instructor will obviously have a careful eye on it all the time. In the early days, the student is concentrating so hard on all the other aspects of the flight that he often forgets about fuel management, so we try to get them into a routine of regularly checking. They should also check the oxygen supply, that the engine is operating properly, their location, and so on. This is all briefed for the sortie, and also during mass briefs, where we get all the students together and go through all the basic points of a forthcoming exercise.

The BFTS course sets a series of specific aims for each training flight. For example, a typical sortie could include demonstrations of the way in which the aircraft controls are used, how the ailerons affect the roll rate of the aircraft, how the rudder controls the yaw, and how the elevators (the tailplane) affect the pitch, pointing the aircraft upwards or downwards. The initial training was (and still is) very simple, teaching the student how the Jet Provost 'feels' to the hands and feet of the pilot. Continual practice ensures that skills are learned well. Even the act of getting the aircraft off the ground and into the air requires a good deal of practice:

You will have to show the guy how to taxi out and line up on the runway, with all the appropriate lookout, and the use of the RT. He'll have to call for take-off clearance, get on to the runway, get the power on slowly, up to ninety per cent, and then make sure he remains stationary at ninety, as that's the brake check. Then up to full power, brakes off, rolling down the runway, keeping it straight initially with the brakes, and then with just the rudder's input once there's enough air speed. By 85 knots you begin to increase the back pressure on the control column, and the nose wheel should be off at about that speed, so the aircraft is just settled on the main wheels, and then at 95 knots the aircraft will fly off. Then there's the gear and flaps to be raised, engine checks to do on the climb, and more RT to leave the airfield circuit … and so it goes on. Further on in the course you get them to change speed, because on the earlier exercises they will have been flying at just a set speed. Now they have to fly straight and level at varying speeds, accelerating and decelerating. Climbing and descending and circuit turns, normally done at height away from the airfield … and then we come back and maybe do a couple in the actual circuit, at the end of the sortie. Then we have a look at stalling the aircraft, how to recognise a stall and how to recover from it with minimum height loss. If the student gets into a low speed situation when he's close to the ground, he's got to be able to recognise the stall symptoms and be able to make recovery action. The main thing to look for is a high nose angle or a low decreasing airspeed. In a stall the controls will become very light and sloppy, and there will be airframe buffet too. We teach them to recover clean, that means with no gear or flaps, and also how to do it in dirty configuration with everything down, putting on the power, overshooting, and raising the gear and flaps. Then there is some continuous circuit-bashing, doing roller landings, talking to air traffic, flying co-ordinated turns on to finals, lining up on the runway at the right place … just building up their confidence. When they have reached a safe standard, we will fly a Dual-to-Solo, flying with the student, then after half an hour we land, get out, and the chap goes off on his own. You have to satisfy yourself that he's done about three circuits safely … they don't have to be incredibly accurate, but the main thing is safety all the time and you have to make a judgement as to whether they're capable of flying the aircraft safely on their own. It's all very subjective, and it's up to the QFI to judge when the student is ready to go solo.

For the student's first solo, he'll do just a quick circuit, taking off, into the circuit, turn downwind, approach and land, and that's his first solo done. After that we do more consolidation, building in glide circuits. Should he have an engine failure, he should be able to glide back to the airfield. For a forced landing we would set up a simulated engine failure away from the airfield, then glide in, achieving a position called High Key, gliding to the Low Key position at the end of the downwind leg at fifteen hundred feet, then turn in for final approach. The next major stage in the training syllabus is known as Sector Reconnaissance. This is the first time that the student pilot is briefed to fly solo away from the home airfield, to gain a degree of familiarity with the local area surrounding it. The pilot is given a simple route to fly over the local countryside, using basic navigation techniques, and after flying the route accompanied by a QFI in another Jet Provost (at a height of 3,000 feet) he then re-flies the same course on his own. After that phase, the student moves on to more advanced stuff, like flying step turns, flying at sixty degrees angle of bank, spinning the aircraft, learning how to position himself for entering a spin, how to recognise spin symptoms, and how to recover. The students will not spin solo, at least not intentionally, but when they're doing aerobatics later in the course and things go wrong, it's possible to enter a spin, so they need to know how to get out of it.

Then comes the solo General Handling flight. Obviously monitoring a solo is difficult. You know what the guy is like, and you've given him a thorough briefing before he goes off, and explained all the rules and regulations that he has to obey. You just expect him to go and do as briefed. He's trying to pass the course so it's up to him to go and practise before he gets to this stage. There's no pressure though, the solo is just his opportunity to practise without an instructor sitting next to him. After this comes an introduction to instrument flying, followed by aerobatics … initially the basic five, which are loop, barrel roll, roll off the top, stall turn, slow roll … and also a wing-over which counts as an aerobatic manoeuvre. We encourage them to practise all these as a sequence, linking them together. If the student doesn't successfully complete a specific part of the training course, he will be allocated additional hours of flying training. A small amount of extra flying instruction can be given, when required, to compensate for occasions when the student encounters difficulty. Sometimes though, it may be that the novice pilot is simply having a bad day, and that's something which can be expected occasionally, and it's a condition with which the instructors are very familiar. On the Spin Aerobatics test, the student will organise an hour's sortie, plan it, and then fly with an instructor, to see if he can safely depart the circuit, do turns, stalls, aerobatics, a spin and recovery, recover from unusual positions, and fly some circuits back over the airfield, doing a radar recovery. If they fail and then do some extra hours, they can then have another go at the test. If they fail that one, there will be progress reports written on the chap, and he may go to Review Action, which means that we keep a special eye on him. He'll maybe get a more experienced instructor, and maybe even more extra hours to help him. We find that people encounter problems at various stages on the course, but if there's a specific exercise where he isn't coping, we have what we call flex-hours, which are used as required. The Gate is simply one of a series of tests, essentially to weed the guys out … with an off-squadron examiner sitting next to the student, with the accompanying additional pressure. The guy has to present his sortie to the examiner. However, assuming that they do pass that gate, they then do more solo, practice aerobatics, navigation at medium level, then a long period of instrument flying, building up their capabilities so that they can fly around with a lower cloud base. Then there are UPs, or Unusual Positions, in which the instructor induces an excessive angle of bank or a high rate of descent while the student isn't looking, which simulates what would happen if the student had looked into the cockpit at some switch, and stopped scanning the instruments properly … He looks up and sees that the aircraft isn't in straight and level flight, so he has to take the correct recovery actions. That all leads up to a basic instrument flying grading, allowing him to fly in slightly poorer weather conditions.

When the students have flown around fifty hours in the Jet Provost, the course has reached the halfway stage. Their course now begins to incorporate more advanced aspects of training, such as the use of various radio navigational aids, and low-level flying at 250ft. The capability to fly at low-level with confidence and safety takes time and some thorough demonstrations from the instructors. The skill of flying so low has to be developed so that the student can judge height visually without depending on the altimeter, which cannot be relied on in any case at low level because of pressure variations. The instrument flying aspects of training are also built upon further, so that the student is capable of flying to a diversion airfield if necessary, bringing in instrument approaches, and Precision Approach Radar, working towards the next rating in the course:

Then it's more maximum rate turns, flying the aircraft on the buffet, to the maximum point of lift. The buffet comes on when you're just approaching a high speed stall, the maximum amount of lift that the aircraft can achieve. If you go into a turn at 220 knots, pull into light buffet at about five-g, the speed will decrease and stabilise at about 150 knots and 2g or so ... and you maintain that on the buffet. If you over-pull, you go into a high speed stall, but to recover from that you just have to release the back pressure on the stick. Further down the course comes a full low-level navigation sortie, more instrument work, and then the Advanced Flying Test, which allows the students to descend through a greater depth of cloud, and fly approaches under Surveillance Radar control. We then combine high level navigation with the low level navigation, and do a mixed profile Navex or Navigation Exercise. The Basic Handling Test at the end of the course involves the student showing the examiner just about every aspect he has been taught, and then he is ready to go on to the next stage of training.

At this stage, the student would be role-selected and, depending upon abilities and preferences, would be streamed into one of three groups. Group One, which included students destined for fast jets, would result in flying another sixty hours on Jet Provosts

RIGHT: A magnificent view of the 'Macaws' aerobatic team, comprising four Jet Provosts from the College of Air Warfare based at RAF Manby in Lincolnshire. Note the very appropriate macaw artwork on the aircraft's noses, and the CAW badge on the fuselage. (*Tim McLelland collection*)

before moving on to advanced flying in the Gnat or Hunter (both replaced by the Hawk in the late 1970s). Group Two students flew another fifteen hours before moving to multi-engine training in the Varsity, until that aircraft was ultimately replaced by the Jetstream. Group Three students were transferred to helicopters, although these students were often selected at an earlier stage of the course. As explained, however, the Group One fighter/bomber hopefuls continued their training at the BFTS:

Having completed the basic course, the students will go on to night flying, night navigation and practice diversions. The night flying skills are similar to those required for instrument flying, but there are stages where you have external references to keep the aircraft in the right place. The students have to develop this ability, as they tend to be staring at the instruments when they could be looking out, but at other times it can be very disorientating if you're looking out ... maybe there are stars, and pinpoints of light on the ground, so you can't tell where you are. There follows a solid period of basic handling and instrument flying, prior to flying an Instrument Rating Test, which qualifies the students to use the airfield Instrument Landing System [ILS]. The speed of low-level navigation flights is increased to 300 knots, instead of 240 knots, which the students have previously been accustomed to. A lengthy period of formation flying is then undertaken, some twelve sorties, using two or three aircraft combinations. We fly basic positioning, breaking away, re-joining, and basic turns. As their competence improves, they do higher angles of bank, and more severe manoeuvres. Tail chasing, formation circuits, approaches and landings, plus formation flying in cloud. After that phase we return to low-level navigation, normally on detachment to places like Brawdy, Leeming, Kinloss or West Freugh, somewhere close to hilly terrain, so that they can practise valley flying, contour flying, leading to a Final Navigation Test. Before that they will also have done some landaways, going to another base, doing a turn-around and flying a different route back. On the test they will have a target included on the route both inbound and outbound, to which they will have to locate the IP – Initial Point – and fly down it, over the target. After that there is more General Handling revision, leading up to the Final Handling Test which represents the very end of the course.

Deciding which type of operational aircraft the student would be best suited to was always a difficult task for the instructors, particularly when the RAF operated such a diverse range of aircraft types when compared to the modern era. There were no specific rules to dictate which students would be better suited to fast jets, multi-engine types or helicopters. The decision was subjective, relying on the instructor's experience:

We have to ask ourselves what the guy is really like, whether he is switched on with the right sort of temperament. For fast jets, we have to decide if he is aggressive enough, whether he gets on with the job, how he reacts under pressure. For example, if his instrument flying is good but he can't fly too well at low level, he might well still make sound decisions, just maybe needing a little more time to cope with the situation. In that case he would probably go to Group Two. If he can fly at low level but can't handle the aircraft too well, he may be better suited to helicopters, where he will be re-taught to handle that type of machine. We always take note of personal preference too. We don't push people to join Group One, but we try to get people to go that way, because that is the type of pilot the RAF needs most of all. However, if the guy has a definite preference for something else, he will probably go on that route unless there is a practical reason why he cannot. In between the flying exercises, the students would also spend a great deal of time studying in preparation for their next exercise on their course. General Service Training (GST), first introduced at (Junior Officer Training) JOT, was also still continued, albeit on a reduced scale. Ceremonial drill was still practised too as was fieldcraft ... learning how to survive in potentially hostile territory. Every course features a Landex, a week-long, outward-bound style course including long-distance walks, making a shelter, and finding food. The capability to survive alone is important for wartime operations, but even during normal peacetime training there are many unpopulated areas of the UK where a pilot could, in an emergency, find himself ejecting into adverse surroundings.

The Jet Provost BFTS course was long and full. The students had a great deal to learn, and many simply did not make it to the end. Problems could be connected with almost any aspect of flying; simple co-ordination of control column and throttle was a common pitfall. Airmanship, remembering checks, navigation, formation keeping, night flying, low-level flying, circuit flying – all were potential areas for failure. Every aspect had to be mastered with competence, and even for the most able of students it was very hard work. A typical post-exercise Jet Provost debriefing illustrates the problems:

Right then, the taxiing first of all … don't stamp on the brakes, keep going in a nice straight line. You must learn to anticipate before the ground marshaller. Would you agree with that?

Okay … Now, the take-off … the line-up was done nicely, but at take-off speed it was obvious that things were not going well. Try to keep the ailerons neutral unless there's a wind on one side. You can play around to find a neutral position … look out of the front a bit more. Notice the way it slopes down … in fact you'll find that the ailerons will work on the ground, against the oleos, so try to get familiar with that. You don't want such a nose-high attitude … When we got off the ground we started to waver, and yet you don't do that on your rollers…

The debrief continues, with the instructor raising many points for the student to consider and act on the next time he flies:

On the manoeuvring I reckoned it had to be better than last time, as we didn't get into heavy buffet. Now, on this bit of navigation: How did we get Harrogate after Lincoln? … yep, just be aware of those navigation issues … remember you don't have to slam the throttle, and if you try rolling while stalling, the aircraft will depart. Don't hold it in the buffet for too long … yes, I know we didn't get the nose drop, but why bother? Why wait for the nose to drop? Look, what's the number-one symptom of a stall? Yes, heavy buffet, so why wait for the nose to drop? … Now, this navigation issue …

You will have to sit down with a map, put some radials on it from Pole Hill and get familiar with it … so what does Pocklington look like? No, I mean without looking at your maps … No, we were never anywhere near Pickering … you said Pocklington … oh, you meant to say Pickering … What can you do with these chaps, eh?! … What was the rate of descent there? It was nearly off the clock, and we don't want to be like that at two thousand feet or we'll impact. Now, back here, point your aircraft at the dead side of the runway, not Leeds … and keep looking out of the cockpit, because the Cessna pilot isn't going to see us until it's too late. And then on finals you have got to cut power, otherwise we won't go down. Don't start searching for the runway, just cut the power, check, hold, and wait … The taxi in was okay. Well, overall the sortie wasn't too bad; you can't expect everything to come together at once … Is that a fair account of what went on? … Okay.

The debrief could sometimes sound pretty brutal, but the relationship between the instructor and student always required plain speaking, and there was never any reluctance to address difficulties as directly as possible. It was in this way that the RAF trained countless combat pilots on the little Jet Provost – an aircraft that never possessed any glamour but was undoubtedly a hugely successful trainer aircraft. Its replacement by the turboprop-powered Tucano was seen by many as a retrograde step, but whatever the arguments over the pros and cons of jet-versus-turboprop trainers, the RAF had to say goodbye to the Jet Provost, simply because the aircraft was exhausted. It had to be retired. Having earned the respect and affection of generations of RAF pilots, it was inevitable that more than a few Jet Provosts were purchased by civilian buyers, and even today the sight and sound of the Jet Provost is not particularly uncommon, even though many aircraft enjoyed only short lives in civilian hands, due to the surprisingly high costs of operating and maintaining them. But the little Jet Provost is still around, to remind us of a long-gone era when the RAF worked hard to produce an endless supply of combat pilots, destined for the ranks of the RAF's Cold War Order of Battle.

2

GETTING A BRAZILIAN

The much-loved Jet Provost was ultimately replaced in RAF service by the Shorts Tucano. It's fair to say that the Tucano has never earned the respect or affection that was bestowed upon the Jet Provost, but it's also fair to say that the Jet Provost was a hard act to follow. For many, the change to a propeller-driven aircraft seemed like a backward step, but despite being powered by a turboprop engine, the Tucano performs as well as the Jet Provost in most respects. More importantly, it does so at far lower cost, and with a cockpit that is designed for a modern environment, replicating the kind of instrument layout that the student will encounter in the Hawk – the next aircraft in the RAF's training process. The Tucano is a functional and unremarkable machine, but it handles well, and is a worthy successor to the Jet Provost. This chapter describes precisely how the Tucano is operated, explaining the details of the training process as outlined by the RAF's specifications. The information and details are based on official training procedures, in accordance with the input of the RAF's flying instructors who apply the procedures on a daily basis:

Learning to taxi the Tucano is a fairly simple process that is easily mastered. However, as with almost every aspect of military aircraft operations, there are pitfalls for the unwary student. Taxiing skills are not taught in isolation from other flying exercises, and the essential techniques will be acquired during early training sorties. Subsequently, the instructor will demonstrate how to cope with specific situations as and when they arise. Although taxiing an aircraft is a relatively simple task, the effect of wind flow on the fuselage, together with the use of differential braking and reverse thrust, may produce unfamiliar effects. Strong crosswinds will affect the directional control. Because of the Tucano's long fuselage and large fin, the aircraft tends to behave like a weathercock turning into the wind; for example, a strong crosswind from the left will encourage the Tucano's nose to swing left. Strong winds will also deflect control surfaces, causing them to bang against their stops, which could possibly damage them, so the control column should be held firmly into one corner, to prevent damage. Tailwinds may trigger an automatic stall warning, which can be cancelled by using the Stall Warning Isolating Switch.

Unlike some aircraft (and most road vehicles that the student will be familiar with) the Tucano is equipped with differential brakes. Each main wheel brake caliper is operated individually by the respective toe brake. Equal foot pressure on the brakes will slow the aircraft in a straight line, whereas unequal toe brake pressure will cause the aircraft to turn in the direction of the more heavily braked wheel. But differential braking increases stress on the undercarriage components and carries the risk of damaging the tyres, and should therefore only be used when necessary. The Tucano has a small amount of available reverse thrust which varies in effectiveness depending upon engine RPM and airspeed, but even at low taxiing speeds and with the engine set at 70% RPM, it will offer some useful additional control. Reverse thrust should be used to slow the aircraft before wheel braking is applied, as necessary.

Taxiways are normally marked by a broken white line, painted down the centreline. These markings are only intended to denote the exact position of the centreline, and while it is generally good policy to follow the line in a small aircraft like the Tucano, the centreline markings take no account of the mainwheel track of larger aircraft, and consequently bigger aircraft could leave the taxiway on tight corners, by following the painted line. Additional markings are applied to the Aircraft Servicing Platform (ASP) to assist with positioning, and departure from these lines may reduce clearances from ground equipment. The captain of the aircraft is always responsible for the safety of his aircraft, and it is his responsibility to avoid collision with other aircraft and obstacles when taxiing, even when under the direction of a ground marshaller. If there is any doubt about adequate clearance from any obstacles, the captain should stop and request assistance. If there is still any doubt, the aircraft should be shut down, and towed to a less-restricted manoeuvring area. The Tucano's wings extend 17ft on each side of the fuselage, and this should be borne in mind.

RT clearance must be obtained from Air Traffic Control (ATC) before an aircraft is allowed to taxi. Taxiing aircraft have right of way over vehicles and pedestrians, but not over other towed aircraft. A good lookout is important before crossing a runway, and a runway should never be crossed before obtaining permission from ATC. The Tucano's engine is very powerful, and excessive use of power will produce rapid accelerations, additionally creating substantial propwash, which can cause damage to other aircraft, or endanger personnel. Taxiing speed should be limited to a fast walking pace in confined areas, on bends, at night and on wet/icy taxiways. On longer, straight sections in dry conditions it is safe to taxi at faster speeds, but it should be borne in mind that undue use of the wheel brakes will cause higher than normal brake disc and tyre temperatures, which may significantly increase the stopping distance required when aborting a take-off.

The Tucano is equipped with conventional aerodynamic controls. In the early stages of basic flying training it is convenient to assume that each control will work in only one axis, but in later sorties it will be demonstrated that both the rudder and aileron produce further effects. Four forces act on an aircraft in flight, these being: thrust, drag, lift and weight. In level and unaccelerated flight these forces are balanced, with weight and lift acting against each other, as do thrust and drag. However, the balance between these coupled forces will vary depending upon the centre of gravity, the amount of thrust and drag. Although the forces can be arranged to balance each other, there will normally be a residual imbalance, causing either a nose-up or nose-down pitching moment.

To keep the aircraft from pitching down, the pilot must pull back on the control column to obtain up-elevator, and the reverse also applies. Naturally a continual force applied on the stick will eventually become uncomfortable for the pilot to hold, so trimming devices are fitted which enable the pilot to select the desired elevator position without using the stick. The trim switch operates a small aerofoil tab, which serves to 'fly' the elevator at the chosen deflection, thus avoiding the need to displace the control column. When the control forces in each axis are reduced to zero in this way, the aircraft is 'in trim'. The Tucano has trim tabs for all three axes. Aileron and elevator trim are driven by a single twin-axis switch on top of the control column, the rudder trim being operated by a switch on the forward face of the throttle.

Pushing the throttle forward increases the engine power, coarsening the pitch of the propeller and causing a corresponding increase in thrust. Changing thrust will upset the balance of forces acting on the aircraft, requiring it to be re-trimmed. Unlike a jet aircraft, power changes in the Tucano also produce changes in directional trim. The slipstream produced by the propeller rotates helically around the fuselage, eventually hitting the port face of the fin, producing a yawing movement. As airspeed is increased, the helical flow elongates, thus striking the fin at a smaller angle, which reduces the yawing effect. Also at higher airspeed the fin becomes an increasingly effective stabiliser, which reduces the yawing tendency still further. Therefore the strongest yawing movements are produced when applying full power at low speed.

ABOVE: Tucano ZF144, resplendent in a striking air display paint scheme. The RAF traditionally provides a solo aerobatics display aircraft and crew for every UK display season, demonstrating to the public the aircraft's agility. (*Tim McLelland*)

ABOVE: An early morning scene on the flight line at Linton-on-Ouse as Tucano crews prepare to embark on a formation flying exercise. (*Tim McLelland*)

LOOKOUT AND 'FEEL'

Military pilots rely on the 'see and be seen' principle to avoid collisions in VMC (Visual Meteorological Conditions). As most military aircraft are camouflaged, visual detection can be particularly difficult at times, and a thorough lookout is always essential, even when concentrating on other tasks. For ease of reporting, pilots use a clock code system to relay information about visual contacts, relating the contact's position to a relative location on a clock-face, assuming that straight ahead is 12 o'clock, 90 degrees to the right is 3 o'clock, and so on.

In addition to lookout, it is important regularly to monitor the position and condition of the aircraft. The appropriate checks are listed in the Tucano Flight Reference Cards carried on each flight by the student, and are known by a mnemonic, FOEEL. A FOEEL check should be carried out roughly every ten minutes:

Fuel. Check the contents remaining, and the fuel flow. This will establish how much longer the sortie can be flown before returning to base. Check fuel balance, and if an imbalance is noticeable, the fuel tanks are trimmed by selecting both fuel pumps to ON on the heavier side of the aircraft, and leaving one pump ON on the lighter side.

Oxygen. The student should check his oxygen contents, the two connections, the flow, and check that the instructor's oxygen indicator is still blinking.

Engine. Check that all indications are within limits.

Electrics. As specified.

Location. Should be fixed using visual features if possible, or by using Tacan and a kneepad map. Care should be taken when using a kneepad map, as if only one minute is taken to fix location, the aircraft will have covered 3 miles without the pilot having looked out of the aircraft.

STRAIGHT AND LEVEL

The visual attitude required to fly straight and level at 180 knots remains constant. If the aircraft accelerates from 180 knots to 220 knots, keeping the visual attitude constant, the wing will produce more lift. The weight of the Tucano hasn't changed and so the aircraft will climb; in order to stay level, the aircraft's AOA must be reduced, to compensate. As the aircraft accelerates, the nose must be progressively lowered, and the reverse applies when decelerating. As IAS (Indicated Air Speed) increases, so does the effectiveness of the fin and rudder, and where relatively large rudder deflections are required to maintain a straight path at low speeds and high power, the rudder deflection should be reduced at higher speeds.

TURNS

Newton stated that a body in a state of uniform motion, travelling in a straight line, will continue to do so until compelled to change by an applied force. So, to turn the aircraft, a force must be applied to it, this being lift. Of course, lift is a force which acts at right angles to the relative airflow, and increasing lift while flying straight and level will simply cause the aircraft to climb. To turn, the lift vector must be inclined, to enable some of it to act in the horizontal plane. However, if the lift vector is simply inclined, the vertical component would no longer be sufficient to maintain level flight, and the aircraft would descend. Therefore, the total lift must be increased, by increasing the angle of attack of the wing, i.e. moving the control column aft. So for a successful level turn, both bank and back pressure is required on the stick.

FUEL ECONOMY

Since power is produced by burning fuel, it is fair to say that the slower the aircraft flies, the longer it can remain airborne. However, as a compromise between fuel economy and aircraft manoeuvrability, most training sorties are flown at 180 knots, which also equates to an easy-to-calculate 3 miles per minute. There are occasions when the aircraft has to be flown for endurance, for example, when asked to 'hold off' by ATC, while another aircraft or emergency is dealt with. In this case the aircraft speed should be reduced to 110–115 knots, minimising throttle movements and flying as gently as possible, to reduce fuel consumption.

EFFECT OF LANDING GEAR AND AIRBRAKE

Extending the flaps, airbrake or landing gear into the aircraft slipstream will cause additional lift and/or drag. This in turn will cause an immediate change in trim due to the changes in aerodynamic forces acting on the aircraft. The extra drag will then cause the aircraft to decelerate, again upsetting the balance of forces, causing a longer-term trim change. In practice, the operation of ancillary services is normally accompanied by an intentional change in airspeed. For example, when the undercarriage is lowered downwind to land at 140 knots, the increase in drag is deliberately used to reduce airspeed to 115 knots. This change in IAS will change the overall trim, and so the transient out-of-trim forces are not trimmed out, being held on the control column, and the longer-term change is then corrected by trimming when established at the revised airspeed. In short, the immediate effect of extending the airbrake is to cause a nose-up change in pitch (the opposite also applying). Extending the landing gear causes a nose-up and then a nose-down change, the reverse being the case when raising the undercarriage.

EFFECT OF FLAP

Lowering the flaps will alter both the lift and drag of the aircraft, which again will result in trim changes. Lowering MID flap produces a relatively large increase in lift for a relatively small drag penalty, and an accompanying nose-up change in pitch. Full flap (DOWN) will produce a large drag increase, a small increase in lift, and a nose-down pitch change. Because MID flap creates lift, the aircraft will climb if the pitch attitude is not modified. As the IAS subsequently reduces and lift decreases, the nose must then be raised to maintain level flight. Selecting flap in the Tucano will always increase lift, so a lower nose attitude will always be required for level flight with flap extended.

CLIMBING

In order to maintain a climb at any given IAS, more power must be provided than for the same IAS in level flight. While in level flight the thrust is only required to overcome drag; in a climb the thrust has to overcome the drag and lift the weight of the aircraft. The power necessary to overcome drag is referred to as 'Power Required', and the amount of power produced by the aircraft's propulsion system is known as the 'Power Available'. If the power available exceeds the power required the excess power can be used to climb the aircraft, and the greater the excess, the faster the rate of climb. In the Tucano the maximum excess power occurs close to the minimum drag speed, so the best rate of climb in the Tucano is achieved at speeds around 125 knots, i.e., relatively slow. Because the Tucano has a great deal of excess power, the aircraft would need to be flown at an AOA of about 15 degrees nose-up in order to take advantage of the best climb speed, climbing so steeply that the day-to-day training altitude would be reached while still in the airfield vicinity. Naturally general handling sorties are not flown close to the airfield, so there would be little point in making such a steep climb. Normal sorties will be flown at a climb-out speed of 170 knots, which will give a greater horizontal separation from the airfield, and, because of the lower nose attitude, will improve the student's forward visibility during the climb.

DESCENDING

As explained, excess power will enable the aircraft to climb, the mechanical energy of the engine being converted to potential energy. If, for whatever reason, the power available becomes a smaller amount than the power required, the aircraft will naturally decelerate, so if the pilot wishes to maintain a given IAS he must convert potential energy to kinetic energy, by descending. It follows that the greater the deficit in energy, the faster potential energy has to be converted, requiring a higher rate of descent. So, for any given IAS and drag configuration, the power setting will dictate the rate of descent. A standard visual descent in the Tucano is flown at 180 knots, with the throttle set at FLT IDLE, and the airbrake in. This is a good compromise configuration, balancing the need to descend quickly and the need to make headway back to base.

FIRAC CHECKS

Before beginning any descent it is important to establish that the aircraft is properly prepared, and FIRAC checks are designed to cover the appropriate points, and should be carried out every time a recovery to any airfield is made, or before a descent to low level:

Fuel. Ensure that there is sufficient fuel for recovery to base. If there isn't enough fuel, this is the best time to find out, because the aircraft burns less fuel at altitude, and the recovery can be modified to save fuel, or a diversion to another airfield can be initiated.

Instruments. Just about every descent will involve the possibility of flying into cloud, so it is important that the student check that both attitude indicators are erect, and that the compass is erect and synchronised. Entering cloud with unreliable instruments is almost guaranteed to create a disaster.

Radio. The appropriate radio frequency for recovery should be set.

Altimeters. Unless a clearance down to a flight level is anticipated, both altimeters should be set to the QFE reading. Before descending, the two altimeters should be cross-checked to ensure that neither is giving an erroneous reading. If there is a discrepancy, the altimeter with the lowest reading should be used.

Conditioning. A rapid descent from the cold air at altitude to the warmer and moister air at lower levels will cause rapid misting of the canopy. After a prolonged period of flying at altitude, the cockpit should be warmed before descending, to prevent misting.

Mis-setting or misreading of an altimeter, or failure to notice a mis-reading altimeter during descent, will greatly increase the possibility of colliding with the ground. Altimeters should be cross-checked at least every 5,000ft in descents that are above 10,000ft, and at least every 2,000ft at any lower altitudes. It is also considered good practice for students to call out the height to which he has been cleared every time the altimeter is checked. This will reduce the chances of inadvertently descending through the cleared height, and, as an additional safeguard, IMC (Instrument Met Conditions) descents below 2,000ft should be restricted to a maximum descent rate of 1,000ft per minute.

CLIMBING TURNS

As previously explained, the rate of climb depends upon the amount of excess power available. The Tucano has a considerable amount of excess power, which can be both a help or a hindrance. When climbing away from the airfield circuit, the reserves of excess power can be used to good effect, gaining height quickly. However, the climb is made to only 1,000ft in the circuit, and climbing at full power would generate a higher rate of climb than is necessary. Consequently, once the undercarriage and flaps have started to retract after take-off, the power is reduced to 60 per cent Tq, which provides sufficient excess power to climb away safely and quickly. Because the downwind leg of the circuit is flown at 140 knots (below the landing gear's limiting speed), the upwind turn is also made at this speed, and this combination of speed and power produces an unfamiliar climbing attitude, which the instructor will demonstrate to the student.

An ideal climbing turn in the airfield circuit will cover precisely 180 degrees in 500ft of climb. Temperature, wind and flying accuracy will affect this aim however, and sometimes the downwind heading will be reached before attaining 1,000ft, and conversely, 1,000ft is sometimes reached before the downwind heading. In the case of the latter situation, power can be used to make a smooth transition from a climbing turn to a level turn. Reducing torque to the point where available power equals the power required, and by lowering the nose, the aircraft can be levelled off, while still holding 45 degrees of bank. This level turn is then maintained until the downwind leg is reached.

DESCENT TECHNIQUES

Regardless of the aircraft configuration, there are two methods of controlling descent: The aircraft attitude is used to maintain a constant airspeed, while the power is used to control the rate of descent (the flight path); and the attitude is used to fly the appropriate flight path, and the power is used to control the airspeed.

Both techniques can be and are used in the Tucano. The former method (attitude for airspeed) is normally used when there is no specific visual aiming-point for the descent (usually the finals turn), and most descents are flown using this technique. However, the latter method (attitude for flight path) is used when there is a specific visual aiming-point for the descent, such as on a visual approach to the runway when landing. It is important to understand both techniques, and when each should be used.

As previously explained, if the power available is less than the power required, the aircraft must descend if the speed is to be maintained. Students are initially taught to descend with the throttle set at FLT IDLE. Because closing the throttle reduces the power available to a minimum, a high rate of descent has to be flown in order to maintain the speed, but it is dangerous to fly such high rates of descent while in the airfield circuit, especially during the finals turn with landing gear and flaps clown. At this stage the rate of descent is controlled with power, never by using the airbrake. With some power applied, less potential energy needs to be converted to kinetic energy, and a higher nose attitude will maintain the desired speed, thus decreasing the rate of descent. The rate of descent round the finals turn can be accurately controlled in this manner. By extending the flaps, undercarriage and airbrake, drag is increased, but the propeller also creates drag. At roughly 10 per cent

ABOVE: ZF345 at the moment of take-off, with a student pilot at the controls and a QFI occupying the aircraft's rear seat. (*Royal Air Force*)

torque (Tq) the propeller will generate zero thrust. Below 10 per cent the propeller will generate drag. Reducing Tq to below 10 per cent can cause a rapid decrease in speed, if a constant attitude is held. Equally, a rapid increase in the rate of descent will be created if the speed is kept constant, and naturally neither situation is desirable when flying the finals turn at low speed in a high-drag configuration. The instructor will demonstrate the very high sink rates, which can occur in this situation, emphasising why it is so dangerous to throttle back to FLT IDLE on the finals turn.

STALLING – CLEAN CONFIGURATION

Lift is generated by a pressure differential of the airflow over the wing. This pressure differential will increase as the AOA of the wing is increased. The Tucano's wing is designed to operate over a range of angles of attack, roughly –4 degrees to +15 degrees, which is more than sufficient for typical training operations. However, as the AOA approaches 15 degrees, the point at which the airflow separates from the wing begins to move forward from the wing's trailing edge. Eventually, as the AOA is increased, the separated airflow will be felt through the control column as 'buffet', caused by the disrupted airflow passing over the tail surface. The airflow becomes increasingly separated, but still generates lift, until reaching a point at which lift begins to decrease, and this is the Stalling Angle. Increasing the AOA beyond this will cause most of the airflow to separate from the wing, causing a rapid loss of lift, and the aircraft starts to sink. Further increases in AOA will simply produce heavier buffet. The transition between an attached airflow and the complete separation of airflow occurs within a very narrow band of AOA, as little as 1 degree. This is a significant factor, which affects the characteristics of each stall. Ideally, the Tucano's two wings would be perfectly identical in every respect, stalling at precisely the same angle as each other, causing the nose to drop with the wings remaining level, but manufacturing tolerances do not incorporate such precision, and so the wings are not completely identical and do not stall at precisely the same angle as each other. As one wing stalls before the other, the aircraft will roll towards the stalled wing, and naturally the roll could be in either direction. Attempting to correct the roll by using aileron inputs will only exacerbate the problem, as down aileron will increase the ACA of the stalled wing, further increasing the asymmetry in lift, and increasing the rate of roll. Consequently the wings should be unstalled before making any aileron inputs.

STALL SYMPTOMS

Closing the throttle and maintaining straight and level flight will cause speed to decrease. This in itself is a warning that the stalling angle is approaching. As the speed decreases further the nose is raised progressively in order to maintain level flight. The ACA will increase steadily, and as the speed reduces, the controls will begin to feel less effective. Finally, the audio stall warning device will sound, and shortly afterwards light buffet will be felt through the control column. The basic symptoms of an approaching stall, therefore, are:

A low speed which is still reducing.
A high nose attitude.
An increasing angle of attack.

Decreasing control effectiveness.
Audio stall warning.
Light buffet felt through the control column.

If all these warnings are ignored and an attempt is made to hold level flight, the aircraft will enter a full stall. The first indication will be the increase in light buffet to a more pronounced heavy buffet, felt through the whole airframe. At the stall the nose will drop and the aircraft will sink, possibly rolling left or right, while the audio stall warning continues to sound. The symptoms of a full stall, therefore, are:

Heavy buffet felt through the airframe and controls.
Nose drop.
Sink.
Possible wing drop in either direction.
Audio warning sound in the cockpit.

During the BFTS course the Tucano is deliberately stalled in order to practise the stall recovery procedure. All stall recoveries are made on the assumption of a worst-case scenario, where the aircraft is close to the ground. The aim, therefore, is to recover from the stall with the minimum of height loss, as there is little point in successfully unstalling an aircraft if it crashes into the ground during the procedure. Without doubt the safest recovery is one made before the onset of a full stall. The first priority of a recovery is to reduce the AOA. Moving the control column forward will stop the buffeting, reducing the AOA below the stalling angle. If there is wing drop, it will immediately reduce, and having successfully unstalled the wings, the next priority is to increase the airspeed so as to avoid re-entering the stall.

Lowering the nose will trade height for speed, or alternatively an increase in power will cause the aircraft to accelerate. The second action, therefore, is to apply full power; to save time and height loss the power is simultaneously applied as the stick is moved forward. In order to ensure that all of the available lift can be used effectively, the wings are levelled, and as the aircraft accelerates the nose is raised to ease out of the descent. It is important to remember that there will be some powerful and continually changing trim forces during the stall recovery. Finally, full power checks should be made when safely established in a climb. Collectively these actions are known as the Standard Stall Recovery, summarised as:

Control column is held centrally forward until buffet stops.
Apply full power simultaneously with stick input.
Level the wings.
Ease out from the descent.
Trim the aircraft.
Complete full power checks when established in a climb.

Should the stall be approached inadvertently at any time, there is no point in waiting for all stall symptoms to appear, and at the onset of any of the described symptoms the stall recovery action should be initiated.

HASELL CHECKS

Before manoeuvring the Tucano in all three planes, or before any exercises that will involve reduced manoeuvrability or a temporary loss of control, essential checks must be carried out. These checks will ensure that, whenever such manoeuvres are flown, they are made in a fully serviceable aircraft which is correctly configured, and that the aircraft is in clear airspace with sufficient recovery height in case something should go wrong. These checks are collectively known by the mnemonic (HASELL) and are:

Height. There must be sufficient height to recover before reaching 5,000ft when flying dual, or 8,000ft when flying solo. The aircraft must also be at least 3,000ft from any cloud. For stalling exercises, 2,000ft above the minimum height should be sufficient for safe recovery.

Airframe. The flaps and landing gear should be in the appropriate position for the manoeuvre, and for a clean configuration stall they should both be selected UP. The airbrake should be tested as it may be required.

Security. Check that the harness is both tight and locked. All loose articles should be stowed. Map bin lids should be secured, pockets should be fastened, and a check should be made that there are no pens stowed unfastened in the external holders on the flying suit.

Engine. The RPM, EGT oil pressure and temperature must all be within acceptable limits. Check that there is sufficient fuel, and that the fuel contents do not fall outside the balance limits for the aircraft.

Location. Fix the aircraft position by using radio aids, or by a visual check. Ensure that the aircraft is clear of any active airfields, built-up areas or controlled airspace.

Lookout. Ensure that the aircraft is clear of cloud by at least 3,000ft vertically, and that there are no other aircraft in the vicinity.

STALLING – APPROACH CONFIGURATION

Having learned to recover the Tucano from an incipient and fully developed stall with landing gear and flaps selected UP, the next stage is to achieve the same capability with the aircraft set up in the approach configuration, with landing gear and flaps selected DOWN. The instructor will demonstrate to the student what will happen if the Tucano stalls fully in the finals turn, and the rapid roll and height loss should convince the student that the symptoms of an inadvertent approaching stall should not be ignored under any circumstances. Recovery should be made upon recognition of the first symptom. There are five major factors that affect a stall:

Weight. More lift will be needed to maintain level flight, if extra weight is carried. Consequently a higher airspeed will be required at any AOA, including the stalling angle, and therefore, the weight of the aircraft will increase the basic stalling speed.

Flaps. At a given AOA the flaps will increase the lift, which is generated by the wings, and so the speed necessary to maintain level flight will be reduced, thus lowering the stalling speed. Extended flaps will also increase the airframe buffet, which could mask the less-severe pre-stall buffet, and the slight asymmetry of extended flaps will increase the likelihood of wing drop at the stall.

Thrust. If the aircraft should stall with power applied, the line of thrust will be inclined upwards as the AOA increases, and a component of the thrust will act vertically upwards, supporting some of the aircraft weight. The propwash flowing over the wings may also cause some increase in lift, and the combined effect is to cause the aircraft to stall at a lower speed. The reduction in effective AOA over the wing's inboard section also means that the stall will occur at a higher nose-up angle.

Icing and Airframe Damage. Damage to the aerofoils, or accumulations of ice, may cause an increase in the stalling speed, by altering the effective cross-section of the aerofoil, thus reducing its lifting characteristics, or causing an early disruption of the airflow over the wing. Of course there will also be an increase in the aircraft weight in the case of icing accumulations.

Loading. All manoeuvres, apart from a pure roll, will require a centripetal force, provided by increasing the load factor, and having exactly the same effect as increasing the weight of the aircraft. The stalling speed will therefore increase when 'g' is applied.

As only a small margin exists between the onset of buffet and a fully developed stall in the Tucano, particularly with gear and flaps extended, two independent artificial stall warning systems are incorporated into the aircraft. An AOA vane provides information for an AOA gauge, AOA indexer and the AOA audio warning. Additionally, the stall warning vane provides information for the stick shaker facility, and both the audio warning and stick shaker operate roughly 4 to 6 knots above the stalling speed. It should be remembered that the stick shaker only operates when the gear and flaps are down. If the AOA and stall vane heaters are not switched on, or if they fail, the systems may fail in icing conditions. Consequently it is important to switch on the AOA/STALL heaters after take-off.

CIRCUITS

Air Traffic Control is responsible for the control of all aircraft movements on the airfield and in the airfield circuit area. Within the airfield's Military Air Traffic Zone (MATZ), all ATC instructions are mandatory and must be obeyed, unless the aircraft captain considers that the action would endanger the aircraft. Local ATC is normally divided between air operations and ground operations, each with an individual radio frequency and controller. The engine start and taxi will be the responsibility of a ground controller, whereas take-off, landing and circuit flying will be under a local controller. The circuit is divided into three sectors: the upwind leg; the downwind leg; the finals turn and finals approach.

ALLOCATION OF LANDING PRIORITY

It is important to make R/T calls when in the correct geographical position, since the order of landing priority is made according to the reported position rather than by the physical position in the airfield circuit. An aircraft that has called 'finals' will have priority over an aircraft still flying downwind. Likewise an aircraft calling 'downwind' has priority over aircraft upwind. If there is more than one aircraft in any sector, the one that calls first is assumed to be in front and has priority. At a BFTS, an aircraft making an instrument approach reaches what is the equivalent of finals at 3 miles from touchdown. As the aircraft will be controlled on a separate frequency, the final clearance to land will not be heard by other aircraft in the circuit, therefore details will be relayed by the local controller.

TERMINOLOGY

Go-Around. Discontinuing a circuit at any stage is referred to as a 'go-around', and although a final approach can still be made, no attempt should be made to either land or roll down the runway. The go-around should be initiated at 200ft or more. Continuing to make an approach below this height will cause the runway caravan controller to fire a red Very flare. 'Overshoot' has the same meaning.
Overshoot/Undershoot. The former describes an approach that would end in a landing beyond the designated touchdown point. The latter refers to a landing that would occur before the touchdown point.
Roller Landing. Describes a landing that is immediately followed by another take-off, without stopping on the runway. Speed does not normally fall below 10–15 knots of the rotating speed at any stage,

and this is a military equivalent of a civilian 'touch and go' manoeuvre. Threshold Speed. This is the speed that needs to be achieved as the aircraft crosses the runway threshold, prior to round-out and touch down. The aircraft speed must not be allowed to fall below the threshold speed figure before this point.

TORQUE EFFECTS ON THE TAKE-OFF ROLL

Simple laws of physics dictate that every action will produce an equal and opposite reaction. Seen from the cockpit, the propeller will be seen to rotate clockwise, creating a reaction that will encourage the aircraft to roll anti-clockwise. A down-force is created on the port tyre, increasing its rolling resistance on the runway, and compounds the previously described slipstream effect, increasing the swing to the left. On most take-off runs, right rudder will be required to keep the aircraft straight on the runway, and compounds the previously described slipstream effect, increasing swing to the left. On most take-off runs, right rudder will be required to keep the aircraft straight on the runway.

EGT AND TORQUE LIMITS

The Tucano will be either EGT limited or torque limited, depending on ambient temperatures. The maximum permitted EGT of 650 degrees will be reached before 100 per cent torque on warm days, and built-in sensors will adjust the fuel flow so as to maintain the maximum temperature. Consequently 100 per cent torque cannot be achieved on hot summer days, and a longer take-off run should be anticipated.

FACTORS AFFECTING TAKE-OFF RUN LENGTH

There are many factors that will affect the length of any take-off run, the main ones being:

Aircraft Weight. The stalling speed is directly proportional to the weight, and so a heavier aircraft must accelerate to a higher speed in order to achieve sufficient speed for take-off. Increased weight also reduces the rate of acceleration, and both factors will lengthen the run.

Wind Velocity. By making a take-off directly into wind, the aircraft will reach take-off airspeed at a correspondingly lower groundspeed, as the headwind speed is effectively subtracted from the required take-off speed. However, only the direct headwind component of the wind will affect the take-off speed and, for example, a crosswind at 90 degrees to the runway will have no effect. Likewise, an aircraft taking off with a tail wind would require a groundspeed that is faster than the required rotation speed.

Runway Gradient. A runway that slopes upwards will cause the aircraft to accelerate more slowly, creating a longer take-off run. Conversely, a downward slope will shorten the run.

Runway Surface. Snow, heavy rain, or a combination of the two, may retard the acceleration of the aircraft, and even small depths of snow and slush can prevent an aircraft from reaching the required rotation speed.

Flap. The use of MID Flap will increase aircraft lift with only a small drag penalty and enable the aircraft to take off at a lower IAS, thus reducing the take-off run.

Temperature. Increased air temperature will decrease air density, and in these conditions a longer take-off run will be required in order to reach the same take-off speed. Additionally, the increased temperature may reduce the power of the engine and so compound the problem.

Effects of Crosswinds. Crosswinds will produce a 'weathercock' effect upon the aircraft and will additionally cause it to drift sideways. When airborne, the aircraft will drift downwind and should not be allowed to sink back on to the runway, as the sideloads inflicted upon the undercarriage could cause damage.

Effect of Flap on Approach and Landing. Extended flaps will provide extra lift, thus lowering approach and landing speeds, and creating a lower nose attitude. The additional drag created will require more power in order to fly the normal glidepath.

Windshear. Strong winds blowing over uneven ground will produce windshear, a change in windspeed, which varies with height. This can cause sudden losses of airspeed and lift, creating a high sink rate. Large power increases may be required to remedy the situation, and in strong wind conditions the pilot should anticipate windshear, increasing the approach speed as a precaution. Gusty wind conditions are also associated with cumulus clouds, and may produce windshear.

FACTORS AFFECTING THE LENGTH OF LANDING RUN

Groundspeed. The higher the groundspeed the more energy must be dissipated during the landing run. High groundspeeds are caused by flying inaccurate threshold speeds, a lack of headwind component, lack of flaps, a high weight, high outside air temperatures and a high airfield elevation (creating a higher true airspeed).

Braking Technique. When landing the Tucano, the shortest landing run will be achieved by lowering the nose wheel immediately after the main wheels make contact, applying full reverse thrust, and application of maximum brake pressure without locking the wheels. But for normal landings, reverse thrust alone is sufficient, stopping the Tucano in roughly 2,000ft of runway.

Braking Efficiency. Water or ice on the runway will significantly diminish wheel braking efficiency. Likewise, runways that are not treated with a 'friction course' will also reduce braking efficiency, but these factors will not alter the effect of reverse thrust.

Runway Gradient. Contrary to the effect upon take-off, a downward sloping runway will increase the length of landing run.

No Reverse Thrust. If reverse thrust cannot be applied for some reason, the landing roll will be significantly increased, by relying on wheel braking alone.

THE EFFECT OF WIND ON THE CIRCUIT
Wind conditions will affect both groundspeed and track, which in turn will affect the angle of bank required in the turn on to the downwind leg. It will also affect the downwind heading, the angle of bank and power required in the finals turn, and the amount of power required on final approach. Both crosswinds and headwinds are almost always present in the circuit, and the effects of each must be fully understood and the appropriate corrections should be made.

WAKE TURBULENCE
A comparatively light aircraft such as the Tucano will be severely affected by the wake turbulence of even small aircraft such as the Hawk. Minimum separations, as recommended in the Flight Information Handbook, should always be observed.

FORCED LANDINGS
While it is unlikely that any student will experience any difficulties in the Tucano, the possibility of engine failure must always be considered, not least because the Tucano is a single-engine type. The Tucano glides very well, enabling students to practise forced landings regularly. However, in the event of a real engine failure the crew would always retain the option of ejecting from the aircraft, especially if far from a suitable diversion airfield. If a forced landing is attempted, and the pilot should fail to complete the manoeuvre successfully, the decision to eject must be taken while flying within the seat parameters. The pilot will have the clearest indication of likely success as the aircraft turns on to finals, but by this stage the aircraft will be flying with flaps and undercarriage down, causing a relatively high rate of descent. Consequently it could be fatal to eject if the height is lower than one-tenth of the rate of descent, and so the final go/no-go decision should be made at 300ft, ejecting immediately if the landing is adjudged unsuccessful.

FORCES ACTING IN A TURN
In turns of up to 45 degrees of bank, only a slight back pressure is required on the control column, and a small amount of power, in order to maintain level flight. Almost no increase in gforce is felt by the pilot.

But in turns involving a 60-degree AOB or greater, the forces increase significantly. For example, a steady 60-degree turn will register '2g' on the aircraft accelerometer, i.e., twice the normal force of gravity. As the AOB increases, more lift is required and back pressure on the stick must be increased. But this situation cannot be sustained indefinitely, because by the time a 90-degree AOB is reached, there is no vertical lift component at all. Very steep turns also produce a large amount of drag, requiring an increase in power, which must be immediately reduced again after the turn so as to avoid accelerating.

MAXIMUM RATE TURNS

Maximum rate turns are often used in aerial combat as both an offensive and defensive manoeuvre. A tight turn will enable a fighter pilot to bring his sights to bear on an enemy aircraft, whereas a similar tight turn would allow a pilot to avoid a missile attack, if flown accurately. To achieve the best possible rate of turn, the pilot must produce the maximum amount of lift possible, pointing as much of it as possible in the direction of the turn. In the Tucano the maximum coefficient of lift is achieved at an AOA just below the stalling angle, coinciding with the onset of light buffet. Therefore, the three requirements for a maximum rate turn are: a maximum AOB, the maximum lift coefficient (light buffet), and maximum thrust (to achieve maximum speed). Altitude also affects performance because the denser air at lower altitude will enable more thrust to be developed from the engine, and increase the amount of lift produced by the wing. However, at BFTS students are taught to achieve a maximum rate turn at a given height. Another factor to consider is the aircraft strength, and although the Tucano is stressed to withstand 7g, normal usage is limited to 6g in order to avoid inadvertent overstress or increased fatigue of the airframe. Attempting to achieve a rate of turn greater than that permitted by the 7g limit would damage or destroy the structure of the aircraft. Therefore the maximum rate in the Tucano is achieved by applying maximum power, using the maximum possible AOB, and flying at a speed, which gives 5g at the onset of light buffet.

MAXIMUM POSSIBLE RATE TURNS

Despite the Tucano's reserves of power, the thrust available is still unable to balance the increase in drag during a maximum rate turn, and even at full power the speed will drop until thrust balances the drag. At about 160 knots the maximum sustained rate of turn consequently equates to about 4g at 2,000ft (roughly 130 knots and 2g at 5,000ft). As 4g is 1g less than the maximum permissible for a maximum rate turn, the deficit in thrust can be compensated by converting potential energy into kinetic energy, i.e., by descending to maintain speed. This produces the Maximum Possible Rate Turn, which can be sustained for as long as sufficient height remains.

G-INDUCED LOSS OF CONSCIOUSNESS

High levels of gforce will produce unpleasant effects, such as an inability to move one's head or limbs easily, and more serious effects such as blackout and loss of consciousness. Rapid increases in g can cause 'G-LOC', or g-induced loss of consciousness, which lasts for very brief periods, but could obviously jeopardise the safety of the aircraft and crew. If a student does experience G-LOC, he is advised to avoid repeating the manoeuvre which caused it, and to report to the Station Medical Officer after landing, to detail the circumstances of the incident. This is part of an effort to gain a greater understanding of the phenomenon. Much less serious is the gradual onset of 'greyout', a loss of peripheral vision, leading to a temporary complete loss of sight. The greyout threshold varies between different pilots, and even an individual's personal greyout threshold will vary from day to day. Naturally, if the pilot's glimit is lower than the Tucano's, he is placing an artificial limitation on the aircraft's performance, so every student is encouraged to keep his gthreshold as high as possible. gtolerance increases with experience, but tolerance can be decreased by other factors such as illness, lack of physical fitness, hunger, lack of oxygen, fatigue or a hangover. gtolerance can be temporarily increased by tensing one's leg and stomach muscles during application of 'g', and by grunting, hence the sometimes rather odd noises heard on airband radios!

SPINNING

One possible symptom of a fully developed stall is wing drop, which is caused by one wing stalling before the other, creating a rolling tendency towards the stalled wing. If permitted to develop further, the roll can lead into autorotation, a self-sustaining roll, with increasing yaw inputs, created by the differential in drag between the stalled and unstalled wing. If this goes unchecked, a full spin will develop, in which the aircraft, still stalled, enters a descending spiral, while rolling, pitching and yawing, behaving in a similar fashion to a gyroscope through all three axes. The Tucano's propeller compounds the problem by creating a slipstream with torque and gyroscoping effects of its own. Consequently the throttle should be set at FLT IDLE during full and incipient spin recoveries.

THE INVERTED SPIN

Should the aircraft be mishandled at high negative angles of attack, the Tucano could enter an inverted spin. The direction of a spin is defined by the direction of yaw, the roll and yaw direction being the same in a normal erect spin. However, in an inverted spin the directions of yaw and roll are in opposition to each other, producing a very confusing visual image for the pilot. Prolonged spinning, either erect or inverted, can cause disorientation, a reduction in the pilot's g threshold, and even airsickness. Spinning is never practised over the sea, as the featureless surface would make visual recognition of rotation very difficult, and likewise, visibility has to be good both at entry and recovery height.

RECOVERY FROM THE SPIN

To recover from an incipient spin, the throttle should be set to FLT IDLE, the controls centralised, and when the rotation stops, the wings levelled. If recovery does not take place within three seconds of centralising the controls, a full spin recovery should be initiated, as follows: check height, set throttle to FLT IDLE, check direction of turn needle, apply full rudder opposite to turn needle indication, centralise control column, and when the rotation stops, centralise the rudder. Finally, the wings are levelled and the aircraft is pulled out of the dive.

AEROBATICS

Aerobatics are an important aspect of military flying, teaching students how to handle the aircraft instinctively and fluently, so that the aircraft can be used to its full potential. Aerobatics often involve large changes of speed, attitude and height, and sometimes require correspondingly large or rapid control inputs. It is important to understand how aileron and rudder inputs can have further effects than those already explored, if either or both are applied for long periods of time:

Aileron. An aircraft flying straight and level, being banked by using the aileron, will begin to sideslip towards the lower wing. Directional stability will then cause the aircraft to yaw in the direction of the slip so that while the primary effect of aileron input is roll, the secondary effect created is yaw. This effect is compounded if left unchecked, the yaw causing further roll (because the outside wing is moving faster, creating more lift, which causes even more roll), eventually developing into a spiral descent. Therefore rapid rolls require a rudder input, in the direction of roll.

Rudder. Yawing the aircraft will cause the outside wing to move faster than the inside wing, the former therefore having greater lift, which will cause a roll towards the inside wing. So while the primary effect of rudder is yaw, the secondary effect is roll. Roll rate can be increased or decreased by applying rudder.

EFFECT OF CHANGING AIRSPEED

The effectiveness of the aircraft controls will vary depending on the airspeed, and for any given rate of response, a greater control deflection will be required at a low airspeed than for a higher airspeed. Therefore, in order to maintain a constant rate of roll or pitch when making large changes in airspeed, the deflection of the control surfaces must be varied. As speed increases, the control force necessary to achieve a given control surface deflection must also be increased. For example, during a loop manoeuvre, a light force and large deflection will be required at the top of the loop, where the speed is low, but towards the bottom of the loop a larger force and smaller deflection are necessary.

STALLING SPEED DURING AEROBATICS

Because the aircraft's acceleration changes constantly during aerobatics, so does the stalling speed, and the pilot must rely on the 'feel' of the aircraft to judge his margin above the stall. In theory, the stalling speed at zero 'g' would also be zero, whereas at the bottom of a loop, for

ABOVE: A sprightly gaggle of nine Tucanos assembling for a flypast over Linton-on-Ouse to mark the successful completion of another basic flying training course. (*Tim McLelland*)

example, where 4g or 5g may be applied, the aircraft may be close to stalling speed, and the most reliable indication is a light buffet. If the aircraft enters heavy buffet the back pressure on the control column should be relaxed, to unstall the wings. When performing aerobatics at height there is no point in taking full stall recovery action at the onset of heavy buffet, but if the aircraft should roll while in a heavy buffet the ailerons should not be used to stop the roll. The roll is an indication that the Tucano has entered an incipient spin, and so the controls should be centralised, and the throttle set to FLT IDLE, in order to recover, after which the manoeuvre can be continued.

EFFECT OF TORQUE

As previously explained, directional trim changes as power and airspeed change, and because acrobatic manoeuvres involve variations in both inputs, the rudder will be required in order to keep the aircraft in balance. Flying out of balance will allow the aircraft to wander off line.

EFFECT OF ALTITUDE

Indicated Air Speed (IAS) and True Air Speed (TAS) are naturally the same at sea level, but the IAS/TAS ratio increases in proportion to height, until at 40,000ft TAS is double the IAS. At the same time, the force required to produce any given change of flight path will vary with TAS, whereas the amount of force that can be produced (the lift) varies with IAS. The result is effectively a reduced rate of turn or pitch, and an increased radius of turn, for any given altitude and gloading, as altitude increases. At altitude, the decreased air density will reduce the air mass flow through the engine, causing a reduction in available torque. This will limit the amount of IAS available, in turn restricting the amount of lift that can be produced. As most manoeuvres will rely on an increase in lift, manoeuvrability is further reduced.

MISHANDLING

During the early attempts at aerobatic flying, the student is likely to mishandle the aircraft, sometimes temporarily losing control, and naturally it is important to take the correct recovery action quickly, in order to regain control and avoid height loss. The more likely mishandling situations are:

g-Stall. Caused by pulling back too severely on the control column at any speed. As soon as buffeting is felt, the stick back pressure should be relaxed, gently re-applying pressure after the wings have unstalled.

Incipient Spin. Ignoring buffeting and continuing to maintain a heavy back pressure on the stick will lead to autorotation, and at this stage it is important to centralise the controls and set the throttle to FLT IDLE.

Vertical Recovery. Control can sometimes be lost in or near a vertical climb, often as a result of entering a loop or vertical manoeuvre too slowly. With a low airspeed recovery cannot be immediately achieved and the pilot should try to push or pull the aircraft from absolute vertical. Throttle is set to FLT IDLE and the controls are centralised until the nose drops below the horizon. A firm grip on the controls is necessary, to counter the high 'snatch' loads sometimes encountered as the aircraft falls through the vertical plane.

Inverted Recovery. If the aircraft is in an inverted position with the nose low, the natural reaction is to pull back on the control column to recover, but this would cause a large height loss. The correct recovery is to roll erect and then pitch to straight and level as quickly as the IAS will allow.

VISUAL NAVIGATION

If a line is drawn on a map, linking point 'A' with point 'B', and a protractor is used to measure the precise heading, and the time taken to get there is calculated, it would be possible, on a calm day, to go from point 'A' to point 'B' exactly on time, precisely along the line as drawn on the map. There would be no need to look at the ground during the flight. If point 'B' were not reached on this theoretical flight, there could only be four reasons why: the planning was incorrect, the aircraft was blown off course by wind, the compass in the aircraft was faulty, or the pilot didn't fly accurately. In practice, the failure to stick on track is usually caused by instrument inaccuracy or sloppy planning, but inaccurate flying and the wind factor will always produce navigational errors on every flight.

Planning. Time invested in map preparation and sortie planning is invariably repaid by confidence in the air. If the student knows that the headings drawn on the map are precise, and knows that the timing marks are correct to within five seconds, he is much more likely to adhere to the intended route, even if he is temporarily unsure of his position. Rushed planning will result in a corresponding lack of precision being carried into the air, and when things start to go wrong the first instrument to mistrust is the map; from that point on the sortie is almost guaranteed to fail. Each student is responsible for allocating sufficient time to plan each sortie, so there is no excuse for bad planning.

Wind. There is hardly ever a completely windless day, therefore, it would be foolish to follow the still-air heading on the map without expecting to deviate from the path. Even a relatively light 10 knots wind abeam the aircraft is sufficient to position the aircraft 2 miles off track after flying a 60-mile leg at 240 knots. In practice there is no such thing as an ideal, windless day. Before flying, the forecast wind for the chosen sortie altitude should be checked. The following simple rules will provide an approximate drift-corrected heading and ground speed for each leg:

1. First calculate the maximum drift by dividing the wind speed by the TAS, in miles per minute (which will normally be 4 miles per minute). A 20-knot wind will, therefore, produce a maximum drift of 5 degrees.

2. Component of maximum drift should be calculated by using the clock-code system as described previously. For example, sixty minutes represents all the drift, forty-five minutes represents three-quarters, thirty minutes equals half, and so on. So a 20-knot headwind at 45 degrees to track will produce a drift of roughly 4 degrees.

3. The head/tailwind component is calculated by subtracting the wind angle from 90 degrees, and then using this figure in the clock-code system to determine the proportion of wind to be used. For example, a 20-knot headwind at 45 degrees from the nose represents a headwind component of 15 knots. So aircraft speed

LEFT: Tucano ZF143 illustrates the huge one-piece canopy that houses the aircraft's two occupants. The canopy hinges sideways and, as can be seen, detonating cords are moulded into the Plexiglas, designed to shatter the canopy if an ejection sequence is initiated. (*Tim McLelland*)

should be increased by 15 knots in order to maintain 240 knots groundspeed. The revised figure should be written on the map.

Compass Error. The Tucano's compass is very accurate, but it could give misleading information if it is not synchronised or is faulty. The best way to ensure against such a possibility is to check the compass continually during the sortie.

Accurate Flying. With accurate instruments, careful planning and an accurate wind estimation, there can only be one other possible cause of an unplanned deviation from track, and that is inaccurate flying. The success of each sortie ultimately depends upon the student's accuracy of flying, and consequently most of the time in the air should be spent equally divided between looking out of the cockpit for other aircraft, and looking at the instruments to ensure that the speed, heading and height are correct. Too much time spent looking at the map will result in track deviations, as will too much time spent looking at the ground. The map should be ignored until it is needed.

The Stop-Watch and the Ground. Surprisingly, it is quite difficult to navigate by making constant reference to features on the ground. Trying to identify every feature, so that there is no doubt about the precise position of the aircraft at every moment, is a recipe for disaster. Only one misidentification of a ground feature will trigger a whole chain of misconceptions until one is completely lost. Indeed there is no point in knowing exactly where the aircraft is at all times, because the provision of a suitable number of navigational fixes is included in the planned route, and if the sortie is planned properly the aircraft will arrive at the prescribed place at the correct time. A stop-watch is a vital navigational tool. For example, at six minutes on the stop-watch, the aircraft will be at the corresponding six-minute mark on the map. And so, as each navigational fix approaches a glance at the stop-watch will give a time that can be matched against the map's timed track, and from this the surrounding ground features can be determined, which should be visible from the cockpit. Timing, however, is important; for example, picking up the map at one minute before the fix point means that the aircraft is still 4 miles away.

The Event Technique. Too many fixes mean too much time spent looking at the map, stop-watch and ground. Too few fixes means uncertainty as to whether drift and speed calculations are accurate. A balance must be achieved, by using the Event Technique. During pre-flight planning, all events that will require specific actions (fixes, turning-points, fuel checks, radio calls, etc.) should be planned to be accomplished at sensible intervals along the planned track. The planned time for each event should be marked on the map. While radio calls and fuel checks should be carried out at the time marked on the map, the fixes and turning-points must be considered at one minute before the event, when the pilot should refresh his memory as to what he should expect to see outside the aircraft. At low level the events should be separated by four or five minutes, whereas at medium level a break of six minutes between events is advisable.

LOW FLYING CONSIDERATIONS

As everyone concerned with military flying knows, modern radar systems can identify aircraft operating at any altitude other than at very low level. Therefore, in an effort to penetrate enemy defences, it is necessary to fly at extremely low altitudes. Practising to fly at 250ft AGL (Above Ground Level) is a good compromise between the conflicting requirements of peace and war. This height is sufficiently close to the ground to emphasise the problems of navigation and handling in this environment, but is also high enough to avoid flight safety problems, and noise annoyance to the general public. Disturbing the general population unnecessarily is always to be avoided, and the following guidelines are to be observed:

Never plan to overfly towns and villages.
Avoid overflying towns and villages below 2,000ft AGL.
Avoid repeatedly flying over the same areas, during a single sortie, or series of sorties.
Avoid overflying large concentrations of livestock, especially at sensitive times such as the lambing season.

Naturally it is not always possible to strictly observe these rules because of the high-speed environment, and from inside the aircraft it is easy to

ABOVE: Doing what the Tucano does best, ZF142 performs aerobatics over a layer of cloud, high above North Yorkshire. (*Tim McLelland*)

forget about the effect that the aircraft will have on the people below. The habit of minimising disturbance is always emphasised.

LOW FLYING REGULATIONS

Because the safety margins are significantly reduced at low level, many lives have been lost during the course of low flying practice. Regulations have consequently been developed that will help to protect students from making the same mistakes others have made in the past, with fatal consequences. Poor discipline at low level is considered inexcusable, and it is recognised as the swiftest way to end a student's flying career. Service aircraft are considered to be 'low flying' when at 2,000ft AGL or less in the case of all fixed-wing aircraft, or less than 500ft AGL in the case of light propeller-driven aircraft such as the Tucano, and helicopters. Low flying is prohibited unless specifically authorised, apart from during take-off and landing, being compelled to fly low due to bad weather, being directed to do so by Air Traffic Control, or during a search and rescue duty. A pilot who has flown at a height less than 2,000ft without special authorisation is expected to record the fact on an Authorisation Sheet.

TURBULENCE

Turbulence can sometimes be severe at low level, especially when the ambient wind speed is greater than 20 knots, over hilly terrain. Instantaneous 'bumps' of up to +4g or –3g can be encountered.

CLOSE FORMATION

By the time a student reaches an operational squadron he will be able to make a rendezvous with a tanker, at night or in cloud, lead similar aircraft through instrument recoveries, and be proficient in all aspects of formation flying. Such a high standard of formation skill requires a great deal of instruction and practice, which is why a great deal of emphasis is placed on such skills throughout flying training:

Formation Positions. The two most common close formations used by front line squadrons are Echelon (for pairs recoveries and departures) and Line Astern (air-to-air refuelling). These formations will be flown extensively during flying training.

Geometry of Turns. Flying in echelon requires an anticipation of the effects produced by two aircraft flying in close formation, each using a different radius of turn. For example, the wingman flying on the outside of the turn will have to climb to maintain the formation references, as the leader applies bank. Additionally, the wingman's lateral displacement results in a larger turning circle, requiring him to accelerate, and maintain a faster speed. The opposite is true for an aircraft on the inside of a turn.

Bank, Heading Changes and Closure. In straight and level flight a wingman applying 1 degree of bank towards his leader (and maintaining it) will produce a continuous turn and an accelerating rate of closure. Alternatively, a one degree change of heading towards the leader will produce a constant rate of closure until corrected. The concept is easy to understand, but students often fail to understand its significance. For example, if an aircraft is too wide of the leader, a small AOB should be applied to alter heading. The bank should then be taken off and opposite bank applied to turn the aircraft parallel to the leader again. Banking towards the leader and keeping the AOB applied until achieving the correct spacing will result in a very dangerous rate of closure with a significant risk of collision.

The Horizon. Formation flying requires the pilot to ignore the horizon, as the only visual references that should be obeyed are the formation data relative to the lead aircraft.

Wing Tip Vortices and Slipstream Effect. The Tucano produces a wingtip vortex, which is capable of rolling a formating aircraft into the leader. It is important to maintain lateral separation between wingtips at all times. In line-astern formation the effect is not a problem, but the slipstream effect will induce a pronounced pitch-up if the formating aircraft flies too high in relation to the leader. The first warning sign of drifting too high is a light buffet felt through the rudder pedals.

Lost Leader in Cloud. Losing sight of the lead aircraft while in cloud is a potentially dangerous situation, as flying close to an unseen aircraft runs a high risk of collision. Likewise, the same situation could result in the student finding himself in an unusual flying attitude, in IMC, when he transfers to his instruments. Visibility problems should be anticipated, and if sight of the leader is lost the wingman must gain separation immediately, before transferring to instruments. The procedure is: Turn 20 degrees, using 20 degrees angle of bank away from the leader, and maintain the heading for 20 seconds; inform the leader of the separation; if the aircraft is in a climb, continue, but if descending, level off; after twenty seconds, resume the leader's original heading and proceed as instructed.

An outside wingman losing sight of the leader in a turn should roll wings level for a minimum of twenty seconds, and proceed as instructed. An inside wingman in a turn should increase bank to 45 degrees. The change of bank in all these situations is very small, and the emphasis is on immediate action rather than severe manoeuvres.

NIGHT FLYING

Contrary to popular belief, night flying is little different from flying in daylight, and in some respects it is easier, because the air tends to be smoother, and there are fewer aircraft airborne. During daytime flying the aircraft will fly with strobes and taxi lights on. At night more attention has to be devoted to their use:

Switch the navigation lights to BRIGHT and have the ground crew check that all three are illuminated.

Leave the navigation lights on while strapping in, as an indication that the aircraft is manned.

Before strapping in, switch the normal interior lighting to ON and adjust the instrument and panel rheostats to give sufficient illumination. Switch emergency lights to DIM and check brightness and direction of each unit before switching the system off.

When the chocks are to be removed, momentarily select the taxi light to TAXI, and then back to OFF. The same signal is used when ready to taxi.

Select the taxi light to AUTO when clearing the ASP, and with the nose pointing away from the ground crew.

When well away from the ASP, adjust the brightness of the cockpit lightning.

The strobe lights should be selected ON, just prior to increasing power to 50 per cent on the runway.

On approach, confirm that the taxi light is selected at AUTO. Landing lights are not essential.

Before entering the ASP, check that strobes, landing lights and taxi lights are all off.

NIGHT VISION

Human eyesight takes roughly thirty minutes to adjust fully to night conditions, although the reverse process takes just one second. No special precautions are needed to achieve full night vision, but the pilot should preserve whatever night acclimatisation he does have by avoiding looking at bright light sources such as the ASP lighting or ground crew torches.

NIGHT FLYING LIMITATIONS

At the BFTS stage in a student's career, night flying exercises are restricted to those considered directly relevant to the learning process, and the following manoeuvres are not permissible at night: Aerobatics (any manoeuvre using more than 90 degrees angle of bank or pitch); practice forced landings; glide circuits; run-in and break over the airfield; low flying (less than 2,000ft).

This detailed look at the RAF's training procedure gives a good insight into how the Tucano is currently operated. It seems likely that the unremarkable (but entirely satisfactory) Tucano will soon be removed from the RAF's inventory as, like the Jet Provost before it, the aircraft is due to be replaced by a more modern (and therefore cheaper) design. But the RAF's training procedures will undoubtedly remain largely unchanged and the procedures outlined in this chapter will doubtless be relevant for many more years to come.

3

HAWK EYED

The Hawk trainer is of course a classic British success story. Sold in substantial export versions around the world, the Hawk even earned the supreme accolade of being selected by the United States Navy as the basis for its new advanced trainer, in the shape of the T45 Goshawk. It joined the ranks of the RAF during the late 1970s, gradually replacing the Gnats and Hunters that were employed in the advanced training at role, based at RAF Valley. The aircraft became responsible for taking students from the basics of flying training (in the Jet Provost) through the world of advanced high-speed flying, until the students were equipped to join an Operational Conversion Unit and begin training on a front-line warplane. As an aviation author, I soon became very familiar with the Hawk, and over the course of many years I was privileged to fly in the aircraft many times until I became so familiar with it that even the 'Red Arrows' team were happy to let me strap in and set up the Hawk's rear cockpit on my own. By the time that I was racing around the skies over Lincolnshire with red-painted Hawks just feet away, either side of me, I had almost forgotten the thrill of my very first flight in a Hawk, so many years previously when the aircraft was still in its early days with the RAF at Valley:

Suitably equipped with flying gear, I join the pre-flight briefing, where the aims of the sortie are outlined in detail and the planned route is examined. Heights, speeds, timings, diversions, weather, call signs, fuel … all important considerations that everybody needs to understand. Naturally a journalist and mere observer doesn't have any major input into the briefing, but as a passenger in the Hawk it is vital to know exactly what is going on and what everybody is going to be doing. With the briefing complete, it's time to collect helmets, life jackets, maps and gloves, and head for the flight line, stopping en route to examine the aircraft's log book and to sign for acceptance of the Hawk for our sortie. Then it's out to the aircraft to conduct a careful walk-round inspection of the airframe before climbing aboard, courtesy of a very substantial and sturdy access platform that is wheeled up to the cockpit side. Attaching oneself to the Martin-Baker Mk 10 ejection seat is a fairly complicated task for anyone unfamiliar with the procedure, but when compared with older ejection seat designs, the Hawk's seat is a relatively simple one because it features a combined parachute and seat harness. Once seated, the first task is to route the leg restraint lanyards carefully through the Drings on the garters attached to one's legs. The lanyards are then plugged into sockets on the base of the seat. If the seat is fired, the lanyards will rapidly retract into the seat, snatching one's legs firmly into the base of the seat to ensure a safe ejection. Flailing limbs would be broken during the necessarily violent exit from the aircraft. Next, I attach the PSP (personal survival pack) clip, ensuring that I am attached to the dinghy and associated assortment of survival aids which are located under the seat cushion. Also at this time, the PEC connection is clicked into place. The PEC (personal equipment connector) connects my oxygen mask tube to the Hawk's on-board oxygen system, ensuring that I have a flow of either pure oxygen or the normal 'airmix' that is routinely supplied. Confirmation that the system is functioning is made by checking a 'doll's-eye' on the instrument panel, which blinks from black to white every time a breath of air is taken. The PEC also supplies high-pressure air to the gsuit, and a test button is pressed to check that the bladders inflate properly. This done, my attention turns to the seat straps, and I thread the two leg loops through Drings in the appropriate lap-straps. The shoulder straps are then routed through the leg loops and pushed into the quick-release box. With the harness fixed, each strap is individually tightened, particular emphasis being placed on the lap-straps, as these will ensure that I remain attached tightly to the seat should the Hawk fly inverted or roll. Another consideration is that a sloppy lap connection will cause quite severe thigh injuries if the ejection sequence is initiated.

This done, the helmet is squeezed into place, the tightening lugs on each side being flipped into their secure position, before lowering the face visor (either clear or tinted can be chosen). The visor is normally lowered before the canopy is closed, to avoid the risk of facial injury, should the MDC (Miniature Detonating Cord) malfunction, and shatter the Perspex. The oxygen mask is clipped on, the radio link is plugged in, and on go the gloves. For a passenger, the pre-flight preparations are then essentially complete, provided that an instructor has already set up the cockpit's switches and dials in advance. In the front cockpit (where the student would normally sit), a 4FTS instructor completes the pre-start checks, before the distinctive exhaust whistle indicates that the Hawk's internal gas turbine starter is now running. The starter is clutched into the main engine mechanism, and the Rolls-Royce/Turbomeca Adour 151 groans and rumbles into life. Further down the flight line, our 'playmate' (another Hawk) is also ready to taxi, and a call over the RT confirms that we are also ready to go. The canopy is closed now, and the MDC safety pins are removed and stowed next to the seat pin, which is also removed, making the Martin-Baker 10B seat live and ready to function instantly if required. The seat has a zero-zero capability, meaning that it could, in theory, be used successfully from a standing start on the taxiway. The rear seat also has a command function that would allow an instructor to eject both himself and then the student, 0.55 seconds later, should this be necessary.

The chocks are pulled away from the wheels, and a slight touch of power pushes us gently forward, before the pilot squeezes the brakes (part of the rudder pedal mechanism) to check their effectiveness. Off forward again, we turn left, along the flight line, passing a long row of parked Hawks, as we head out to the active runway. Moving along the concrete, it quickly becomes clear that the instructor has an excellent all-round view from the Hawk's rear seat, including a superb forward view over his student's head. The differential steering is effective, although it does take time for students to learn to control accurately the Hawk's speedy taxiing capabilities – unchecked, the aircraft could quickly run away. At the holding point, just short of runway 32, we join our sister aircraft, and there is time for the crews to make a last-minute check of everything before lining up on the runway. Air Traffic Control gives 'Snapper One and Snapper Two' clearance to line up and take off, and we take up line-abreast position, either side of the runway centre line markings. As 'Snapper Two' rolls forward, we are briefly buffeted by the Hawk's exhaust, as the sprightly little trainer roars away ahead of us with a perceptible crackle of sound and a burst of brown smoke. Our engine throttle also goes forward, and after checking the engine RPM and TGT (Turbine Gas Temperature), the brakes are released and we surge forward, the pilot guiding the nose towards the centre line with brake pressure, and then with rudder input as the speed increases and the rudder begins to 'bite'. Acceleration is quite brisk, and at 90 knots the control column is gently pulled back, the nose rises, and at just over 120 knots we are airborne, having travelled little more than halfway along the runway. The landing gear is selected 'up', and a couple of muted thumps indicate that the gear doors have closed, with indicators on the instrument panel confirming retraction. Flaps are also raised quickly, before reaching their limiting speed of 200 knots, which is already fast approaching. Still climbing straight ahead, we pass over Holy Island before turning to port and formating alongside 'Snapper One'. The two Hawks continue in formation across the southern side of Anglesey, through the low-level 'one-way' route across the Menai Strait.

Hawks inbound to Valley take the northern side of this corridor, thus easing the congestion which would otherwise occur in this relatively small but busy region of airspace around Valley and the RLG (Relief Landing Ground) at Mona. Once over the mainland we climb through the cloud cover, and break formation, allowing us to continue our solo demonstration flight. During the outbound flight there is time to take a closer look at the Hawk's internal layout. The first impression is that there is a relatively large amount of room in the cockpit, and the instrument panel has a very clean, logical and uncluttered appearance. The majority of non-flying controls are in the front cockpit, making the rear 'office' appear even more tidy, especially as this particular aircraft (XX184) is a standard T Mk 1, without the HUD (Head Up Display) associated with weaponry-equipped Hawks that were used by the RAF's Tactical Weapons Units, and particularly the Sidewinder-capable T Mk 1A that formed part of the RAF's reserve fighter capability. The main attitude indicator dominates the instrument panel, with the compass immediately below it. To the left is a combined speed indicator and to the right the main altimeter. It doesn't take much effort to quickly learn the position and function of all the controls and students become familiar with the aircraft's interior. Clear of the gloomy coastal weather, we make a slow descent over Wales, down into the lush hills, where we settle into the standard low-level altitude of 250ft at a speed of 360 knots, a convenient 6 miles per minute which is also a useful cruising speed for the aircraft. Down 'on the deck' the impression of speed is, to say the least, attention-grabbing. As one pilot describes it, it is 'like driving a high-speed go-cart', and for students fresh from the Tucano, the increase in speed is quite noticeable, especially during the first few Hawk flights. As for the aircraft itself, the instrument panel is very similar to that of the Tucano (intentionally so, as the Tucano's instruments were designed to match the Hawk layout), and apart from the obvious extra speed, the Hawk's handling is also quite similar, although there is a longer delay between the application of power and the accompanying increase in speed. The Tucano's turboprop propeller provides instant power, and students tend to forget the small time-lag associated with a jet engine that needs time to accelerate. Judging the low-level

250ft height takes some practice, but students who have progressed from the Tucano will already be accomplished low-level flyers, and at RAF Valley the main task is to learn how to handle a much faster machine. To a complete novice such as myself it is very difficult to determine the precise altitude, especially in the Welsh hills, where the terrain rises and falls by the second, and although we maintain a fairly steady altitude, it's clear that the distance between us and the ground varies quite rapidly, within certain parameters.

The RAF pilots would, in some circumstances, like to fly lower than 250ft, not least because a realistic wartime operational height has long been regarded as sometimes being only 50ft or even less. Naturally it would be impossible to operate at such heights over the UK, and for training purposes such potentially lethal heights are out of the question; 250ft is regarded as being a fair compromise between the need for realism and the equally important need for safety. Additionally, the nuisance factor has to be considered, and the residents of Britain's countryside could hardly be expected to ignore fast jets flying by at anything less than that altitude. Even at this height, there's no disguising the Hawk's presence as seen (and heard!) from the ground, and every effort is made to spread the low-level training over as many varying routes as possible, so as to avoid creating too much disturbance to any particular location. In fact, the resulting restrictions make the student's low-level map look like a planner's nightmare, littered with hundreds of 'one-way' routes, avoidance areas, restricted zones, and prohibited airspace. The RAF is permitted to fly in the spaces between all these restrictions, and in a crowded island like the UK it makes route planning very, very difficult, even though the RAF instructors know the best areas in which to route the training flights.

For the purposes of this demonstration sortie, we take a low-level route at the speed of 420 knots, representative of the kind of speeds used by students towards the end of the Hawk syllabus. Navigation sorties are flown to airfields all over the UK, ranging from Chivenor in Devon, right up to Lossiemouth in Scotland. Even civilian airfields are sometimes chosen for practice diversions, to places such as Ronaldsway on the Isle of Man, or Liverpool or Blackpool. Certainly the Hawk has sufficient fuel capacity easily to reach any destination required by 4FTS. The first turning point on our route is the southern edge of Llyn Clywedog, a lake running alongside the B4518 road. I am just starting to think about looking for the feature when we turn hard over the lake, having already reached it. Unfortunately for an untrained passenger, it's very easy to forget the speed at which things happen, and one has to learn (quickly) to anticipate the route's ground features as they rapidly unfold. Heading south, we approach Llanidloes, which has to be avoided either laterally or vertically in order to comply with restrictions. We choose to go around the town, carefully negotiating a scattering of avoidance areas and noise-sensitive spots (farms, hospitals, etc.). Low-level flying is thus illustrated as being very demanding, having to cope with restricted airspace, bird hazards, other aircraft, bad weather, and still maintain 250ft, while following a predetermined route, keeping an eye on fuel, talking to air traffic controllers and, in many cases, leading a formation of other Hawks too. To the untrained observer it looks very easy, despite sounding like an impossible task when explained in theory. In reality it is indeed an almost impossible task, at least for a new student.

The next ground feature to look for is a steep hillside on the western edge of Lake Vyrnwy, which serves to confirm that we are still on track. This time, thanks to some warning from my instructor, I manage to see the spot we are looking for, as it seemingly flicks by us at 420 knots, almost within touching distance, or so it seems. Racing onwards, literally bouncing up and down in the turbulence, we approach our IP, our Initial Point, which serves as a 'funnel feature' into a representative target run. Replacing the 1:500,000 scale map on my lap with a larger 1:50,000 chart, we arrive at our IP, which is a spot height on the A543 road at Bryn Trillyn, south-west of Denbigh. By looking at the larger map one can identify the key features which lead the eye towards the selected target, in this case a small bridge on the River Elwy, south of Abergele. The map has already been pen marked with ten-second intervals for me, but despite a running commentary from the front seat, I find it impossible to read the map accurately and simultaneously identify the real features outside the cockpit when they approach at such phenomenal speed. I choose to look directly for the target, and with some verbal assistance I manage

to find it, even though we are virtually overhead by the time I finally establish where we are. My instructor assures me that the skill can be developed, in time, but it's hard to be convinced.

Having visually 'attacked' our target, we make a steep climb out of the bumpy low-level airspace, and head for the crisp blue sky above us, in order to see some other aspects of the Hawk's capability. First of all, an opportunity to sample the Hawk's uncomplicated stalling characteristics, and with a climb rate in excess of 4,000ft per minute, we quickly reach an altitude of 10,000ft, from where we make a careful 360-degree visual search of the surrounding airspace, before starting the demonstration. The throttle is pulled back, and the airspeed slowly bleeds away as our attitude becomes distinctly nose-high. A gentle wing buffet (felt inside the aircraft as if one is riding over a cobbled street) signifies that we are approaching the stall, and suddenly we are stalling, not violently or steeply, but unmistakably heading downwards. With a burst of power and the control column placed centrally forward, the recovery is swift, and the whole manoeuvre is surprisingly unexciting and far less unnerving than it might sound in theory. Back in the early days of the Hawk's developmental flying, the aircraft did possess rather less pleasant stalling characteristics, hence the addition of airflow breaker strips along the Hawk's wing leading edge, which provided a more satisfactory stall warning, and wing fences, which cured a fairly severe wing-drop problem. Stalling demonstration complete, we return to our nominal height of 10,000ft, where the instructor invites me to take the controls. As I am unashamedly in the non-pilot category, the opportunity to 'play' with an aircraft in the Hawk's category is something of a rare treat, although one's initial reaction is (perhaps naturally) to be over-cautious with the controls. It quickly becomes clear that only the smallest of movements on the control column is needed to achieve fairly large pitch and roll changes, and the amount of effort required to achieve a manoeuvre is almost finger-tip, rather than a full arm's-worth of pressure. After some encouragement from the front cockpit, I begin to gain some confidence and try my hand at snap rolls to port and starboard. At first, the overwhelming urge is to close one's eyes and hope for the best, but the gut-churning and totally disorientating effect of high-speed rolls quickly becomes enjoyable, and almost addictive. The instructors are keen to point out that in many respects the Hawk's handling characteristics are very similar to the Tucano's, just a little faster. Likewise, the Tornado is often referred to (at least in terms of handling) as a 'big Hawk'; therefore the Hawk provides an ideal step between the two types. To sample something of the Hawk's turning capability, I am instructed to put the aircraft into a left-hand turn, and with some care I achieve a fairly steady 3g turn. The instructor invites me to pull back on the stick still further, through 4g, and 5g, my gsuit fiercely gripping my legs and stomach as I sink heavily into my seat. My instinct is to ease off in order to stop the torture, but I hear 'a little more, a little more' from my pilot, and then we're at 6g, and still we're turning. The Hawk is stressed to 8g, and 6g is the day-to-day maximum regularly achieved by students. It's a pretty severe amount though, and just a few seconds of sustained turning can lead to loss of vision or even blackout. Consequently, students like to build up a 'gtolerance' as much as possible. Pulling big amounts of gforce is never much fun, but with practice and the right technique (essentially a lot of muscle clenching and grunting), the punishing effects of gforce can be countered, but it's always a relief when it stops.

With this exhilarating but exhausting demonstration complete, it is time to return to Valley, and as we fly through the Menai 'gate', we head for RAF Mona, to sample a demonstration of a forced landing. Simulated engine failures are a regular part of the Tucano and Hawk training schedules, and as we approach the airfield overhead at 5,000ft, the instructor calls the control tower and declares a 'practice pan', cutting power and turning to starboard over the upwind end of the runway at the 'High Key' position. Speed slows to 165 knots, and we enter a descending turn through 4,500ft, heading downwind as the landing gear is extended. Arriving at the 'Low Key' point directly abeam the touchdown point, we're at 3,000ft, still turning and descending, the speed reducing to 150 knots. The flaps are extended, and as the runway slips into our 12 o'clock position, the instructor keeps the Hawk directed at the initial aiming point, roughly a third of the way down the runway. Once satisfied that we're going to reach the runway, the nose is repositioned towards the runway threshold, and we head steeply

down, in a most alarming fashion, the ground seemingly rushing up to meet us. But shortly before touch down we pull out of the descent and flare out the approach, to gently touch down on Mona's runway 23, rolling forward for a few hundred feet before selecting full power again, unsticking, and climbing straight ahead, tucking away the gear and flaps. Turning right, we stabilise at 1,000ft and 360 knots, as Valley's controllers give us permission to join the airfield circuit and land. Descending once again to 500ft, we flash in over the airfield boundary and break hard left in a 4g turn, climbing back to 1,000ft, extending the airbrake, and positioning on to the downwind leg. Speed reduces to 150 knots, and the landing gear is extended once more, with mid-flap selected. Once abeam the touchdown point (the runway threshold), we make a 180-degree descending turn on to finals, while lowering the flaps to the fully extended position. The aircraft is positioned on short final approach to runway 19, at 250ft and 130 knots, gently descending, and touching down at 115 knots. With a fairly stiff breeze almost directly on our nose, the landing run is short, and in a matter of minutes the aircraft is off the runway and back on the flight line, where I reposition the canopy and seat safety-pins in their 'safe' locations before the canopy is opened and a welcome rush of fresh air comes into a hot and sweaty cockpit. I also detach the leg restraints and, more importantly, the PSP connection. It is easy to forget about the PSP, but climbing out of an ejection seat with the connection still in place causes the automatic functioning of a distress beacon which will needlessly (and very embarrassingly) attract the attention of the nearby Search and Rescue Wessex helicopter crews! It would also cost the culprit more than a few drinks in the Mess bar that night! Once the engine has stopped and the remaining seat and PSP connections are unfastened, my helmet is removed, and I climb out of the cockpit, on to the port engine intake, and down the access ladder. For a passenger the flight is over, but for a student a detailed debriefing would follow, tracing every aspect of the sortie, examining what, if anything, went wrong, and what can be improved and built-upon during the next sortie. For the aircraft the day certainly wasn't over, and an hour later XX184 was back in the air with another crew.

ABOVE: The British Aerospace Hawk T Mk 1 first flew in August 1974, the prototype XX154 (illustrated) eventually joining the ranks of the RAF, and the aircraft type commenced its advanced flying training role in 1976 with No 4 FTS at RAF Valley. (*Tim McLelland collection*)

Just a brief taste of Hawk flying is sufficient to indicate that the advanced and tactical flying training syllabus is more than a little challenging. As the instructor says: 'The student has got to be able to do virtually anything at anytime. If we can't generate that ability to think and act flexibly at this stage in their training, then we're wasting our time.' The instructors are able to spot a number of potential faults as they begin to arise, thanks to their collective years of experience. Sometimes it can be a basic handling problem, perhaps being unable physically to control the aircraft properly in close formation. Sometimes it can be fuel awareness that proves the student's downfall:

If he doesn't continually monitor the fuel situation he may not notice if he begins to use fuel faster than planned. He has to make sure that he can 'get home' with sufficient fuel, so he should make an effort

to do something if the problem arises, such as truncating part of a navigational route, or climbing out of a low-level route earlier than planned. Or maybe diverting to another airfield, as even that would mean that the student is at least aware of his fuel state, and has made a positive move to do something about it.

Poor navigation is another important point that the instructors watch out for, as is low flying – that is, flying too low, below the minimum set at 250ft. As a QFI explains: 'Sitting at 250ft, letting the hills rise and fall away beneath him, is just not acceptable. The student is expected to maintain 250ft at all times.' Navigational faults often stem from a tendency to over-read the map. The key to successful visual navigation is, as the instructors explain, 'to be able to put the map down and not look at it'. For example, if there is a leg of perhaps eight minutes to fly, in a straight line, the student is guaranteed to arrive at the end of the leg at a prescribed time, provided that he flies the speed accurately, and there is no significant crosswind. Consequently, the instructor will encourage the student to perform a work cycle, putting the map down, maybe for four minutes, then rechecking his position before continuing, or making speed or directional adjustments as necessary. The important point is to convince the student that putting the map down is not a recipe for disaster, and that he can look out for other aircraft, check fuel, and generally be more aware of what is going on. As one instructor points out: 'Over-reading maps is a common fault, and it's rather like driving down a motorway, wanting to leave at exit twenty, and then checking his map against every exit until he gets there.' On the other hand, he also points out that the Hawk flies much faster than the average car travels down a motorway, and the kind of navigational error that can be achieved in a car in maybe an hour can be achieved in a Hawk in roughly seven minutes. As the course progresses towards tactical weapons operations, the learning curve increases rapidly and life gets even more difficult. The fear of failure is intense, but by this stage the students can at least be fairly confident that, if they do fail, they can probably re-role on multi-engine aircraft or helicopters. Indeed, some pilots go on to multi-engine types only to retrain later as fast-jet pilots, so the change in type doesn't have to be irreversible.

Having mastered the skills necessary to fly the Hawk safely and accurately, the student is now taught how to apply his capabilities by operating the Hawk as a weapons platform rather than as merely a flying machine. Until 1992, the tactical weapons training course was totally separate from the advanced flying training syllabus, and was conducted at RAF Brawdy and Chivenor, within Strike Command. Later, the advanced and weapons courses were combined, split between two identical Flying Training Schools at Valley and Chivenor. The aim, as an instructor explained, is 'to take an ab initio student, or abbos, as we call 'em, and turn them into baby fighter pilots'. The Hawks utilised for weaponry instruction are fully combat capable, able to carry Sidewinder air-to-air missiles under each wing. The Hawks also wear the markings of four reserve air defence squadrons, and in the event of an armed conflict these Hawks would form part of the United Kingdom's air defence, being flown by FTS instructors. One of the first tastes of tactical operations is flying in 'battle formation', a term that becomes familiar from here onwards, through the Operational Conversion Unit and on to an operational squadron. No longer does

ABOVE: The Hawk was designed for use not only as an advanced trainer for the RAF but as a weapons trainer too, enabling students to learn the basics of ordnance delivery and air-to-air combat. XX157 is pictured during development trials, carrying practice bomb dispensers and a centreline Aden cannon pod. (*Tim McLelland collection*)

a formation consist of simply 'hanging on to the leader's wingtip'. The students now begin to separate their aircraft by lateral distances of maybe a mile or more. The first real introduction to weapons operations is what is called a 'Simple Cine' sortie, and is explained by a QFI:

Many of the Hawks are fitted with gun sights, and likewise we also carry a cine camera which will film whatever is seen through that gun sight. We fly a dual sortie with an instructor and a solo sortie, following another Hawk, giving the student a chance to get used to the idea of aiming his gun sight at another aircraft. We then develop the skills into Cine Weave sorties, during which the aircraft being chased starts to manoeuvre more severely, making the task of training the gun sight on to him that much more difficult. At this stage they're obviously not firing any bullets, but they are shooting a roll of film, and after these flights their accuracy will be assessed.

Navigational skills are also developed, and by this stage the student will fly almost every sortie at a speed of 420 knots, making regular IP-to-target runs, formation flights, and many solo sorties, some of which will still be observed by a QFI or QWI (Qualified Weapons Instructor) who will 'chase' the student in a second aircraft. By way of an illustration of how a tactical sortie is flown, I joined a two-aircraft exercise at RAF Chivenor on a low-level sortie around Wales, in a defensive battle formation. The meteorological brief indicated that the conditions would be less than perfect for flying, with broken cloud at a base of 1,500ft and a visibility of about 30km, although over north Cornwall there would be heavy sleet and hail showers. I would take the back seat of XX186 while a QFI would fly from the front seat, acting as a wingman to a student pilot who would lead us along a route that he had planned as part of his routine training. The student's back seat would also be occupied by a Staff Navigator, and this would be the first time that he had flown with anyone other than a QFI, effectively being the student's very first experience of flying with a navigator behind him, as he would do on a front-line squadron. As indicated by the met brief, this would be quite a demanding sortie for the student, as he would be expected to adhere to his planned route as closely as

ABOVE: XX285 is high over Mablethorpe en route to RAF Marham. Based at Leeming, the aircraft was one of No 100 Squadron's aircraft, tasked with training support duties including the provision of target facilities. (*Tim McLelland*)

possible, while navigating around the anticipated bad weather, back on to track. As I was regularly reminded, the clouds over Wales tend to have very solid centres. Our route was intended to take us north from Chivenor, across the Bristol Channel, into Wales, and on into two IP-to-target runs. The student was given the locations of the two targets, and plenty of advice as to how to plan a suitable route, but otherwise everything was left entirely up to him. There are no pre-set routes as such, and the complete exercise, from brief to debrief, would be under the direction of the student.

Strapping in, and pre-start checks are performed as normal, and at 11.40 the two Hawks, call signs 'Kelly One' and 'Kelly Two', are on the threshold of Chivenor's runway 28, ready for a formation take-off in line-abreast formation. With XX352 on our starboard side, the throttles go forward and, with a nod from the leader, the two aircraft surge forward together, and just twelve seconds later we are airborne, climbing straight ahead over the Devon coastline before turning right on to 030 degrees to take us north over the Bristol Channel, at 250ft.

Once out over the expanse of grey water, we assume our tactical formation position, line-abreast, separated by roughly 2,000yd. Over the featureless water it is easy to keep visual contact with the lead Hawk, although at times the separation does vary, and the aircraft becomes little more than a distant black shape, low on the horizon. Speed increases to 420 knots, which equates to 7 miles per minute, an acceptable speed for fuel economy, requiring about 93–94 per cent power setting on the throttle. It would be possible to fly at 450 knots, but this would necessitate full power almost continually to maintain that speed at low level, whereas 420 knots is a more comfortable cruising speed that can easily be maintained.

Approaching the sprawling industrial complexes around Port Talbot, we pass over Mumbles Head, turning due north in a 'shackle' manoeuvre, a high-speed tactical turn, in which the lead aircraft (on our starboard side) passes across our six o'clock position, as we turn left, stabilising on our port side. Reacquiring our sister aircraft was difficult, at least to my untrained eye, despite having climbed to 2,000ft while crossing the Port Talbot area. Back at 250ft, visual tracking is even more challenging. Another minute passes and the gforce piles on again, in a tight 5g shackle turn on to 305 degrees,

ABOVE: The view from a Hawk's rear cockpit, looking forward at the student as he checks his 3 o'clock position. (*Royal Air Force*)

LEFT AND BELOW: The RAF's advanced flying training is conducted at RAF Valley on Anglesey. No 4 FTS comprises two component squadrons, No 208 being one, with No 19 the other (as illustrated), but the latter unit disbanded in 2011 to make way for No 4 Squadron, equipped with the new Hawk Mk 2. (*Royal Air Force*)

making a three-minute run towards our first IP. My attention is constantly shifting between my attempts to keep track of the lead aircraft and my hand-held map, on which the QFI has drawn out the route. Over the IP, the student switches to his 1:50,000 scale Ordnance Survey map, as we race over a small estuary linking Carmarthen with the bay. Just to our left is the Pembrey weapons range, which is used by both 4FTS and 7FTS for air-to-ground attack profiles. Our target is a radio mast close to the A478 road, but I never even see it, being mentally lost in a confusion of rolling landscape, cloud, driving rain and gforce. We turn on to 096 degrees just a few miles east of RAF Brawdy, formerly the home of Chivenor's sister Tactical Weapons Unit. Another tight turn on to 096 degrees takes us east on a five-minute run towards Llandovery, the weather gradually deteriorating, with increasing heavy showers forcing the student to lead us around the worst patches, without straying too far from the intended route. His capacity to quick-think course corrections is vital, and, to complicate matters still further, other low-level aircraft are occasionally seen flashing between the hills in the charcoal grey murk. The seemingly open countryside can be surprisingly crowded at times. We turn hard on to 056 degrees through a long valley feature, with a patch of woodland appearing on the north-eastern side, as predicted by the map. A large hill with two aerial masts is our next turning-point, where we make another 5g turn on to 327 degrees towards our second Initial Point, through another valley, being careful to maintain a respectable separation from a nearby hospital. The target is located in a small village, within a 'basin feature' on the map, and at 420 knots we race overhead, before making another gut-wrenching 5g turn on to 218 degrees.

With the two targets successfully located, our route now takes us back towards Chivenor, still at low level, and to check whether the student is paying careful attention to his wingman, the instructor flying in the front seat of my aircraft decides to sneak in slowly towards the lead's 6 o'clock position. Not surprisingly, we don't get too far before the student notices our deliberate error, and requests us to move out to port again. Also very much on his mind at this stage is the amount of fuel remaining, and at various points in each sortie 'Bingo' fuel calls are made, denoting specific minimum amounts of remaining fuel to reach diversion airfields or home base. The Hawk's excellent range encourages any inexperienced pilot to ignore the fuel status to some degree, but many combat aircraft are much less forgiving in terms of available fuel, so the QFIs try to teach the students healthy habits. The final route turning-point rolls into view, in this case a spot height west of Lampeter, where we set course for the Bristol Channel on 183 degrees, turning on to 177 degrees as we cross the Welsh coastline, and close up our formation. Back at a more intimate distance from XX352, our respective speeds become much more evident, as the aircraft lurch up and down, left and right, in the turbulent low-level air. Earlier in the BFTS course, maintaining close formation would have been a major challenge for any student, whereas by now, on the way back to Chivenor, formation flying is almost taken for granted. We pull up steeply to cross the coastline again near Ilfracombe, and we head for the airfield circuit at 500ft, reducing our speed to 360 knots before we arrive. Sweat-soaked and exhausted, we're soon back on the ground for a detailed debrief.

The tactical training course continues with more weaponry use, introducing the Strafe Phase, in which the student fires the Hawk's 30mm cannon (housed in an under-fuselage pod) at ground targets on the Pembrey range, near Swansea. Four sorties are flown, starting with a dual flight (with a QWI). When the student is adjudged to be safe, he flies a Cine sortie, substituting film for bullets, and this will establish whether he can be trusted to fire live ammunition safely. He will then fly two more sorties using fifty rounds on each, flown against acoustic recording targets at Pembrey. After completion the film will be examined by a QWI, and the score recorded at the range. If either the QWI or range safety officer is not happy with the results, the sortie is 'DNCOed' (Duty Not Carried Out) and is then reflown, emphasising that, even when flying solo, the student is still being constantly monitored.

Later in the syllabus the student is introduced to ACM (Air Combat Manoeuvring), the basic skills of the fighter pilot. Students are given a taste of ACM at RAF Coningsby, a front-line Tornado F3 base, and also the home of the RAF's Air Combat Simulator, a twin-dome computerised facsimile of the view from two Tornado F3 cockpits, in which the students can fight an opponent realistically without leaving the ground. Coningsby's simulator is quite convincing, and from the pilot's seat inside a representative Tornado cockpit the projected 360-degree picture surrounding the cockpit gives a good impression of actually being in the air. The simulated image of a marauding fighter is also most effective, and perhaps the only limiting factor is the fixed altitude of the projected image, which can lead the unwary to concentrate on the outside image, forget the altimeter, and crash.

The simulator gives students a 'feel' for the necessary skills, before taking to the air in their Hawks to fly ACM for real. At this stage, aerial combat is completely new to them, and although they have flown tail-chases, the concept of fighting another aircraft while trying not to get shot down is both confusing and exciting. Initially the student flies roughly 300–500yd behind a target aircraft, and uses the gun camera to simulate firing rounds of ammunition at the target Hawk. Dual sorties with an MI are first flown, and then the student flies solo against an instructor. The emphasis then changes back to air-to-ground operations, flying more sorties over the Pembrey range, this time with practice bombs, progressing to a combined bombing and strafe configuration in which the student makes attacks with both bombs and ammunition on the same sortie. Once their ground-attack skills have been properly learned, the students then fly SAPs (Simulated Attack Profiles) combining the navigation sorties with ground-attack profiles, flying simulated attacks against off-range targets.

For tactical training the Hawks are limited to 6.9g, and although 7g and more wouldn't damage the aircraft, such stress would use up airframe fatigue life rapidly. Negative-g limit is set at –3.5 but as a QFI comments: 'You would have to do some fairly exotic aerobatics to reach that kind of figure, and as we're more concerned with combat, we try to leave that kind of flying to stunt planes.' Chivenor and Valley are well positioned for Hawk operations, with plenty of unrestricted airspace all around, allowing the pilots simply to select areas with the best weather, to conduct the flying exercises. For ACM, the students are not permitted to fly in or even near cloud, because although running away into a bank of cloud might be considered a valid tactic in a real fight, during training it is discouraged. Another restriction is the need to fly over the sea, because of the cannon ammunition that is being expelled. For ground-attack profiles, all off-range targets are predetermined and approved by the QWIs and QFI, and are normally bridges, TV masts, towers, or something equally obvious, and certainly are never private houses. Although the targets may be pre-selected, the students are free to select their own routes to them. By this stage in a student's RAF career, the training costs have been phenomenal, and understandably, the RAF would not want to lose a student after spending in excess of £3 million, unless he really couldn't tackle the job or was simply unsafe in the air. As a QFI explains:

For a couple of years, both at Brawdy and Chivenor, we ran what was known as a streamed syllabus, whereby we found out how many slots would be available for ground attack or air defence, and then gave the appropriate number of students the corresponding course. However, we have now returned to the way we used to do things, giving everybody the same course, and only at the very end do we direct them to specific front-line aircraft. We pushed very hard to go back to the common course because we believed that it produces a better result. For example, you finish up with air defenders who have actually experienced the other side of things, and they have actually flown Simulated Attack Profiles on ground targets, and so on. We're now able to be more flexible, as hitherto we could have streamed people to either mud-moving, or air defence, only to find out later that the chosen route wasn't the one they were best suited to. Now we can wait until the end of the course, the logical time to decide their future, and we can put the round pegs in the round holes. The combined syllabus doesn't cost us much more than the previous one, in terms of flying hours.

Although the aims of the streamed course were admirable, the results were disappointing. The students did receive additional flying time that was devoted to one of the two roles (fighter or ground attack), but they failed to gain an appreciation of the skills employed by their fellow pilots. For example, a fighter pilot learned little about the ground-attack pilot, and in a wartime situation the knowledge of his opponent's limitations could be vital. By experiencing some ground-attack flying, the fighter pilot could learn more about the way in which the bomber pilot, or tank-buster pilot, thinks, and use the knowledge to his advantage. Naturally the reverse applies too, and the ground-attack pilots learn how fighters operate. After a great deal of careful consideration, the Tactical Weapons Units (TWUs) returned to a combined course, and all students learned the same skills, equipping them all with the same degree of understanding and ability, producing a better 'all-round' result.

The TWU instructor (a former Lightning pilot) continues his description of the course:

We operate two squadrons at Chivenor, and at any one time each will have two courses running within it. The course lasts sixteen weeks, and we get a new course starting every four weeks, so by using advanced mathematics you can see that we get through about twelve courses each year, with ninety-six pilots passing through. The first stage of the course is a Convex Phase, intended to ensure that each student can confidently fly the Hawk, rather than simply relying on 'paper' assurances from Valley. It is also important that all the pilots understand the TWU SOPs [Standard Operating Procedures] concerning basic rules of safety, etc. The Convex consists of a dual sortie involving general handling … a solo general handling flight, just to allow the student to become familiar with the local area, and one or maybe two instrument rides. In fact we often fly just one instrument ride which is in effect an Instrument Rating ratification, as they arrive here with a current Instrument Rating.

RIGHT: The Royal Navy also operates the Hawk in the training support role. The Fleet Requirements and Air Direction Unit (FRADU) is based at Culdrose in Cornwall. Four of the unit's hawks are pictured flying over a much-loved local landmark, St Michael's Mount. (*Royal Navy*)

The Convex Phase continues with formation flying:

This begins with four sorties, starting with a dual flight, that is, pilot and instructor together, at medium level, which is then followed by a solo medium-level sortie. We then take them into low-level formation, with another dual, followed by a solo. The first two sorties involve a small amount of close formation, usually a formation take-off, followed by about fifteen minutes of close-up work, with a close formation recovery and landing. We then move to tactical formations, that's a battle formation, which they will have looked at while they were at Valley, but not very thoroughly; and we aim to give them a really good grounding in battle formation flying, as they're going to be using it throughout the course and, indeed, through the rest of their fast-jet flying career.

Once the students have been thoroughly trained in the operations of the Hawk within a tactical environment, the instructors introduce the first part of the weapons phase. The Hawk shows its talons for the first time, as a bomb, gun or missile carrier:

We start with what we call Simple Cine. As you know, we have a gun sight fitted in the TWU Hawks, and we have also installed a camera that can film whatever is seen through the gun sight. With this we will fly a dual sortie and a solo sortie, before we move to Cine Weave, which is slightly more complicated, in that the aircraft which is being chased will perform some rather more dramatic manoeuvres than on the first two sorties, while the student attempts to follow him and keep his gun sight trained on him. They will do one dual flight and then a couple of solos, and the resulting gun-sight films will be assessed. At the end of the flights they will receive a mark, indicating their ability to track another aircraft at the correct range. This is their first introduction to weapons, and although they're not firing anything other than photons, they are taking film.

Although the Cine Weave sorties are assessed, it would be wrong to imply that all the other sorties are not carefully monitored:

Every flight will be written up. If the sorties are dual, the reports will be written by the instructor who was in the aircraft at the time. There are a few sorties that are not written up, however, for pretty obvious reasons. Solo general handling for instance. How could you prepare a report on that? On some of the later solos they will be monitored by an instructor who will follow in another aeroplane, and so in reality very few sorties they fly here are not assessed in some way or other.

Having mastered the techniques of high-speed navigation at low level during the Valley course, the TWU students are expected to develop their skills still further. Looking through a detailed table of the TWU syllabus, the instructor explains:

At about the same time that we introduce Cine, we bring in the navigation phase, our aim being to have what we refer to as weather options. If we're running a low-level phase, we also like to have a medium-level phase at the same time, so that if the weather is too poor at low level, at least we can do something with them, and of course the reverse applies, when we encounter bad weather at medium levels. At the moment, the low-level navigation phase consists of seven sorties. One is just a check-out to ensure that they have all the fundamentals right, and that they can safely fly the Hawk at 250 and 420 knots, these figures being the general minimum height and maximum speed on low-level flying training over the UK. We normally fly sorties at speeds that are multiples of 60 knots, because that's the easiest sort of figure to use for working out timings. In the Jet Provost or Tucano they will have flown at 180 or 240 knots, which is either 3 or 4 miles a minute. They might have gone up to 300 knots too. At Valley they will have started doing their Navexes at 360 knots, which is 6 miles a minute, and they will have finished at 420 knots. Here at the TWU we jump straight in at that speed. If you went to a Tornado or Phantom squadron you would find them flying at that sort of speed during a transit flight. Obviously you could go much faster, but that would use more fuel, and we would also cause even more of a nuisance to people on the ground. You would be using a higher power setting that would create more noise, and we do try very hard indeed not to annoy the general public. In the Hawk a speed of 420 knots is a useful cruising speed, and if we flew any faster we would be rather hard-pressed. We could aim for 450 knots, but we would be using full power almost all of the time, whereas we can maintain 420 up hill and down dale. Typically in our Hawks we need 93 or 94 per cent power to maintain the speed.

The navigation exercises continue, with what the TWU instructors call 'Chase Navs'. For the first time in their careers, the students will fly in company with another aircraft that will not be in formation with them, but will be following their progress. From the vantage point of this chase aircraft, an instructor will watch the student, checking the height and speeds that are being flown:

This puts a little bit of added pressure on the student, knowing that he's being watched all the time. We will carefully check the height he's flying, particularly if he's too low, as we would want to deal with that very quickly. We also see exactly where he goes on his navigation, while he flies a route that lasts about fifty minutes, including a couple of targets to overfly en route. One of the sorties will normally be a land-away, where he will recover the aircraft to another airfield, often near the east coast, at places like Coningsby or Cottesmore. Then we can fly the next sortie on the return to base. Next comes what is really a major progress step, Nav Four, which involves a pair of aircraft, usually with an instructor and student in each aircraft. On this flight the student will fly as a battle formation pair, not only flying the planned route, finding a couple of targets, but also doing the whole sortie in a defensive battle formation. We find that this is a particularly difficult sortie for the students, as they are already working very hard navigating and flying at low level, and now they have the additional worry of manoeuvring a formation successfully, to compound their problems. A good defensive battle formation is important, because it is the only way that the pilot can keep a good lookout to make sure that he's not going to be bounced – that is, attacked by another aircraft. By flying about 2,000 yards apart, line abreast, they can check each other's six o'clock area for any marauding aircraft which might be trying to shoot them down.

On these particular Navexes we don't have anyone trying to bounce the formation, but it allows them to get used to this kind of operation, ready for later sorties when we will launch a bounce aircraft, to see just how good their lookout is. On one exercise, one student will lead the formation until halfway, at which point the lead will be handed to the other student, which allows each student to spend half of the flight leading, and the other half formating.

The next three sorties require the student to lead the flight, without the aid of a pre-planned route. He will be given a target, plus a fairly large amount of advice as to how the route should be planned, but exactly how he reaches the target is effectively his own decision. Once the route is planned, he will be expected to brief the pilot of the second aircraft, and lead the flight. On Exercise Five a staff navigator (a TWU instructor) will fly in the rear seat of the student's Hawk. This will be the first time that the student has ever flown a dual sortie with a 'passenger', someone other than a flying instructor, introducing the practice of flying with a navigator, rather than another pilot, in the back seat. For the first time, the student will be operating his aircraft as a true two-man fighting machine. The pressure continues to build as more responsibility is transferred to the student:

The workload gradually increases. On the days when you have good weather, your workload isn't enormous, as all you have to do is to keep looking well ahead to see where you are going, but in more marginal conditions it can involve a great deal of decision making on the part of the student, as he will obviously have to navigate around the weather. Clouds can have solid centres when you're flying low level.

After completing the navigational exercises, the course returns to more weaponry training, with the introduction of the strafe phase:

In this phase the student fires the Hawk's 30mm cannon at targets on the ground, normally on the weapons range at Pembrey in Wales, not far from Swansea. It consists of four sorties, starting with a dual in company with a QWI, a Qualified Weapons Instructor. Assuming that the student is judged as being safe, he is sent off to fly a Cine sortie, without any live rounds in the gun, simply taking film as if he were actually firing the gun. If that sortie shows that he is safe enough to be let loose with live ammunition, he will fly a further two sorties with real bullets in the cannon. Each sortie will be with fifty rounds, to fire at Pembrey's acoustic target. Upon completion we will record their score, and the film will be debriefed by a weapons instructor. After having seen the film, or if the range safety officer decides he wasn't happy with the way the sortie was flown, the flight could be DNCOed as we say, that means Duty Not Carried Out. If that happens we make them re-fly the sortie, so even though they're flying solo, they are still being constantly monitored.

Even at this advanced stage in the student's career the pressure not to fail is still very high, and some will find that they cannot keep pace with the TWU course:

We don't have a dramatically high chop rate here, and on average it probably works out at about ten per cent. People do get suspended for a number of reasons, first and foremost being our concerns over safety aspects. It's not just that they obviously mustn't be at risk of writing off themselves, anybody else, or the aircraft, but we have to consider that we could squeeze through a student on this course only to have him kill himself and a navigator later on, in a Tornado or Phantom, doing God knows how much damage on the ground as well. So all the time the main point is safety. However, there are some things in the course that some people just can't cope with. They might have performed quite well until now, and then they come to a brick wall, as far as their progress is concerned. I like to think that we're a fairly caring unit, in that we wouldn't simply suspend a student unless we saw no further point in carrying on. We would aim to re-fly a sortie if we felt the potential for success was there. For example, we can look at the navigation phase: suppose that someone performs adequately until he gets to, say, Nav Five – and then got himself hopelessly lost, or flew dangerously, or maybe didn't show any due consideration to his fuel state, that sort of thing … we would probably re-dual him, and fly him on another sortie, with an instructor. We might send him off on another solo, but not if there

were any safety concerns in his performance. If we then considered him safe to be sent off on his own again, he would try and re-fly the original sortie that he failed.

The flexibility and positive nature of the TWU syllabus are reflected by the series of gradual steps that can be taken to bring a student back 'up to par' without suspending him from training:

If, after that stage, he still hadn't got what it takes, we would put him on review, which is a way in which we can highlight the fact that we've got a problem. And this enables us to give the student some more flying hours. We would give him a little remedial package, maybe two dual Navexes or something like that, and during these sorties we would monitor his progress. Throughout we would aim to be pretty flexible, but if by then the student still appeared unlikely to crack it, we would probably chop him at that point. If he did show even a glimmer of potential, we might send him off on another solo, and maybe give him the benefit of the doubt. It's not a rigid, laid-down procedure whereby we say that the student has got maybe two more sorties to make good and then he's out. Every case is treated on its merits.

Returning to the core of the course itself, the training continues:

Also within the basic phase is some night flying. While they are at Chivenor they will normally fly three night trips. They will have done this before at Valley, so we start with a simple dual to check their safety, and then on to a solo, followed by another with plenty of instrument work, runway approaches, a practice diversion to another airfield, maybe a couple of emergencies and so on. We usually fly these exercises spread over two nights. Then, finally in the basic phase, we move on to ACM, or Air Combat Manoeuvring, the basic skills of the fighter pilot. Before this the chaps will have spent some time on the Air Combat Simulator at Coningsby, flying a number of sorties in their twin-dome simulator. This will give them a basic feel for ACM, and it will tend to make their progress rather quicker when they get into the air. We then start them off with two dual sorties to teach them the basic combat manoeuvres.

The students now begin training in the skills of aerial warfare, learning how to become fledgling fighter pilots:

It's important to remember that combat is completely new to them. They are now fighting against another aircraft, trying desperately not to get themselves shot down, and whilst they will have flown tail-chases, this is a whole new concept to them. At this stage we will be simulating gun-armed aeroplanes, positioning the aircraft between three and five hundred yards behind the target aircraft, within the effective range of the Hawk's cannon. Again the use of the camera will simulate the gun, and show the student how he is going to shoot down the other guy. If the first two dual sorties are judged to be safe, they will then fly three solo one-versus-one sorties, fighting with an instructor. Tucked in somewhere in the middle of the course will be what we call a Tac-IF check, where we take the guy up on a dual sortie and throw some emergencies at him, check out his instrument flying capabilities, and make sure he hasn't been developing any bad habits, and so on.

Following the ACM phase, the students return to operations in the air-to-ground role, flying 10-degree dive bombing sorties over the Pembrey range. As ever, the first sortie is flown dual, then with a cine camera, and then two further sorties carrying live practice bombs:

We then progress to what is called Bombing and Strafe, where we not only give them four bombs to play with but fifty rounds of ammunition in the cannon as well, and they will be given a half-hour slot on the range in which to drop the four bombs and fire the fifty rounds. They will be given two such combined sorties. The next flight will be a level bombing sortie flown dual, followed by three 'hot' sorties where they are actually dropping bombs. The practice bombs are released at four hundred knots, at a hundred and fifty feet over the range. However, we never fly below two hundred and fifty feet when we're off the range. I might add when we're flying at two hundred and fifty feet, on the Navexes, the height will be judged visually, as naturally a pressure altimeter is no good whatsoever in that sort of environment, when the terrain varies so much.

Following this phase, another general handling and instrument flying sortie is included, with an instructor in the Hawk's rear seat. There then follows a formation sortie, usually flown with four Hawks, serving to refresh the student's formation-flying skills learned earlier in the course. Two low-level evasion sorties follow (one dual, one solo), in which two aircraft will be flown around a low-level route in battle formation, with a third Hawk acting as 'bounce' aircraft, assuming the role of an 'enemy' fighter out to shoot down the unwary student. The basic aim of the exercise is simply to avoid being shot down (albeit in simulation), but the student is also expected to make 'track progression':

The aim is not to turn off track and fight, but to run away, preferably towards the target. Then we include another ACM sortie, the Hawk being fitted with two AIMnine Sidewinder missiles, to serve as an introduction to flying with missile armament. We can simulate either the Golf or Lima version of the Sidewinder, the Golf being a stern attack version, whereas the Lima has an all-round capability, and could be launched from any angle.

The next part of the TWU course is the SAP phase, the Simulated Attack Profile:

Again we build them up gently, with SAP One being flown dual, showing the student the fundamentals of bombing an off-range target, although naturally we don't actually carry bombs, and we don't fly below two hundred and fifty feet, being off range. After the dual they go off and fly the sortie as a pair, initially with the staff pilot leading, but then with the student in the lead. This works towards the final three SAPs where they will have a bounce aircraft thrown in too, and again the first of these sorties will be flown dual. In addition to the pressure of the bounce aircraft, we also pile on the pressure at the planning stage. Initially, the student will be given plenty of time to plan his route, normally around three hours, and he will be given two targets to locate. However, towards the end of the course the instructors will limit the planning time, giving the student the target locations two and a half hours before the take-off time, effectively giving him just one and a half hours to plan the sortie, thirty or forty minutes to brief it, and twenty minutes to get out to the aircraft and get airborne. Time over target is also introduced, so that the student not only has to locate and overfly the target but also reach the position within a few seconds of a time designated by the instructor.

The continual increase in pressure is designed gradually to improve the student's capabilities:

All these little things build up, so that by the end of the course they are achieving a standard that they just wouldn't have thought possible at the beginning. But every individual skill is important. For example, with time-over-target tasks, you must remember that in an operational situation you may have a target that is the subject of a co-ordinated attack by other RAF squadrons or even other NATO forces. You may simply get a message to be over a specific target at a specific time, and have just a one-minute bracket to get in there and do the job. In the real world there could be four A10s going in two minutes before you, and a recce Phantom running in afterwards, so the timing is very important. For us, we're not actually fighting a war, so it's purely a training asset, of course.

The TWU leads the students gently, slowly introducing different aspects of aerial warfare that become increasingly important as the students move towards the end of the TWU course, and a posting to a conversion unit prior to joining an operational squadron.

Returning to air-to-air combat, the fledgling fighter pilots are taught to test their skills against an airborne banner, towed by one of the TWU Hawks:

It's a sort of hessian flag, which we attach to one of our Hawks which has been modified for target-towing. The banner is towed at a distance of eight hundred feet from the Hawk, and the student goes off with a weapons instructor in the back seat, flying a second sortie with the cine camera, and then flying three hot sorties with ammunition in the gun. Judging how the student scores is relatively simple, as we tow the target back here, dropping it on the airfield near the runway, before bringing it back to the squadron for all to see.

We paint the bullets with a coloured dye, with a different colour for each aircraft, and as the bullet goes through the target it leaves a coloured trace on the hessian. In this way we can put four or even six aircraft on the same target. Moving on, there's the advanced ACM phase, starting with a one-versus-one sortie using missile armament this time, as well as guns.

As with all of the earlier exercises, there is a set sequence of flights, building up from the first sortie. Flying as a pair, the students aim to 'shoot down' a singleton:

Two of these flights are flown dual, and if they perform okay they do the next flight solo. The fact that we do two flights dual in this case indicates that there is a great deal to teach them, with lots of safety aspects to consider when there are three aircraft in the same part of the sky. We then move on to a formation revision trip, and then the Air Defence phase, where we introduce them to practice intercepts, that is, the job that will really be their bread-and-butter on their front-line squadron. There's patently a great deal that we cannot teach them, as our Hawks don't have a radar, but we can give them the basics. We use a fairly close control from a nearby radar unit, starting with things like two-versus-one daytime intercepts, and ending up with Capexes, launching two aircraft on a Combat Air Patrol in a given area, while another bounce aircraft tries to penetrate the CAP. Finally we have an Adex, an Air Defence Exercise, which is normally a simple Capex, but in this case we actually scramble the crew from the crewroom.

These final sorties are very hard work, navigating from one CAP area to another, all the time being threatened by the 'bounce' aircraft. The student has to make sure that he is not in danger of being shot down, and that the 'bounce' Hawk doesn't penetrate the CAP area, acting, as it does, as both an 'enemy' interceptor and bomber, by constantly changing modes as the circumstances dictate:

For something like a medium-level practice intercept, we would fly as a pair under the control of a ground-based radar unit. This would be rather like being in a Tornado or Phantom with a damaged radar, having to rely entirely on directions from the radar controller on the ground. Alternatively we could fly under what we call Bravo Control, whereby there is a supply of radar information from the ground, but do our own control or direction. The basic information is the same, but in the latter situation the student has to decide what to do with the information. We would aim to manoeuvre the pair until we were in a position either to shoot down the intruder, or to perform a visual identification, as we don't automatically assume that the intruder is hostile.

One of the instructing officers sets out the planned procedure:

Air combat manoeuvres are taught right from the first exercise. On ACM One we try to concentrate on the offensive manoeuvres where the student has the upper hand, then on ACM Two we introduce the defensive moves where the student is being attacked. Moves include high yo-yos, low yo-yos, and rolling scissors, which isn't strictly an offensive move but it's one that the Hawk tends to get into quite often. It's really a mutually defensive move, when you're both trying to avoid being shot down. A horizontal scissors is a low-speed move where the two aircraft are constantly turning in towards each other, crossing, and turning back towards each other again. The one who gets himself spat out in front of the other is the one who gets shot down, when the other guy gets his gun sight on to you. In most aircraft, horizontal scissors is quite common, but because of the Hawk's very good low-speed handling we often take the fight into the vertical. This is because you're effectively trying to arrest your forward progress, and a good way to do that is to barrel-roll around. Unfortunately the other guy will do the same thing, so we end up with the two aircraft in a kind of double spiral, with one aircraft on top while the other is on the bottom, while both pilots are trying to get behind each other. If you can't get behind the other guy, and you realise that you're starting to lose, it's important to realise this and run away, sooner rather than later. We try and fly two aircraft that are matched, performance-wise, but obviously if you put up one Hawk with pylons attached to the wings and another that is completely clean, the latter will win, with less drag and more lift at its disposal. However, the skill of the pilot is obviously a major factor.

By the time that the students fly ACM Three or ACM Four, the instructors will have identified the good, the bad, and the average:

The instructor will feed them situations, but the students vary dramatically. By this time you have identified some who are very good, maybe to a standard whereby the instructor just can't give him an inch. We're not going up there simply to shoot down the student, we're trying to bring out the best in him. But you do get some who take to ACM naturally, and others who have to work very hard to keep up. You have students with whom we have to fly a very hard ACM fight, and others who really have barely got a clue, and indeed, after their first solo attempt, you feel that they haven't got what it takes, and then we recommend that they go up for another dual flight.

Safety is a very important aspect of air combat:

We're very keen that the students must not break a minimum height of ten thousand feet, so we're talking about a height of 2 miles up, and that is the ground as far as we are concerned. The student must not be dangerous as regards collisions. There must be absolutely no risk that he's going to hit the other aeroplane. We have to be sure that he's not going to run out of fuel, and that he's going to handle the aeroplane and engine in a safe manner. The Hawk is a very forgiving machine. You can mishandle it to a great extent and it never bites you, whereas an aircraft like the Phantom most certainly will. The Hawk is an excellent trainer for the Tornado, which features almost foolproof handling as well. The Rolls-Royce Adour engine in the Hawk is an incredibly reliable engine, but like any highly tuned unit, it works exceptionally well until you get to its limit, and then it will surge. An engine surge usually occurs because of mishandling the throttle, for instance, in a heavy buffet, turning behind somebody, accelerating the engine, and then hitting the slipstream of the aircraft in front of you. You disrupt the airflow into the intakes to such an extent that the compressor is literally gasping for breath, and too much fuel is flowing through, resulting in a surge. Normally you hear a bang or a loud popping, and certainly the gas turbine temperature dial will rise quite rapidly, but as long as you catch it quickly you can just close the throttle. If that doesn't cure it you have to stop-cock the engine and quickly perform a re-light. I must say that you don't get many instances like that, but you do train the students to be ready for it. We would wish that the aircraft didn't clock up a 7g count, for example, so we set the limit at six-point-nine. Seven wouldn't do the Hawk any damage, but it would use up fatigue life on the airframe. As for negative-g, you would have to do some fairly exotic aerobatic manoeuvres to pull any, but the limit is three-point-five. Other things? Well, we wouldn't be impressed if anyone went into a spin during combat, not deliberately anyway. Chivenor is an ideal place for air combat, as we have virtually free airspace all around us, so we can get airborne and go where the weather is best. In combat you mustn't go into cloud or even go near it, so we need a block of airspace that is completely clear. Running away into cloud might be something you could do in a war, but it isn't something we encourage. Because of the cannon ammunition, ACM is normally flown over the sea, in an area that is confirmed as being clear. For the ground-attack students, the off-range targets are always predetermined features that have been approved, such as bridges, electrical sub-stations, transmission towers, industrial buildings and so on. Certainly they are never private houses. We prefer to have a number of predetermined targets, but the routes are not normally pre-planned, and we let the students plan their own routes, which is much more valuable for training purposes. They plan it and they have got to live with it, and it does mean that the routes do vary, and nobody has to live with a steady stream of aircraft racing over his house all day. We run a very traditional system of assessment here. The rest of the air force has largely gone over to a system approach with objectives, aims and all sorts of things. The net result is that they end up with write-up forms where you have to assess what the guy knew, what he needed to know, did he do this or that, and all finishing with ticks in boxes. I'm totally biased, I'll admit, but what that achieves is

mediocrity. In that system the guys don't have to excel, they merely have to be good enough, and that doesn't exactly encourage you. Here at Chivenor, each sortie is written up by the instructor, and it's purely his assessment that is written down. The length of the write-up will probably depend on how well the student did. If he did well, a couple of sentences will normally suffice. We do give the student a mark from zero to three, with zero being a fail, and a three being something just short of God. So you don't often get a three, and we can amplify the score with a plus or a minus, and the average sortie tends to be a score of two. The scores will be there on the wall for all to see, so we've introduced an element of competition, and the students can take the piss out of each other when they see how they have performed. The write-up is confidential, however, and although they will obviously be thoroughly debriefed after each sortie, they are not encouraged to read the write-ups, even though they can if they want to. They might get the wrong impression if they did, as what tends to get written down are the bad points only. There's little point in writing that the guy did okay, as we don't need to take any action on that kind of performance. However, we don't lock up the books, and human nature being what it is … By encouraging the students to read their reports, I doubt if you would have such an honest system, because we would then write them in the knowledge that the students are going to read them, and so we might be tempted to pull a few punches. I prefer the system we have got now, writing exactly what we think, with an honest debrief, and if he still wants to read his report, we let him. You see, the guy might be a complete wanker, but you would be reluctant to put that on the write-up. You have to be honest but constructive as well. At the end of the day, the student will end up with a couple of folders. One will contain all the sortie write-ups, and the other the weaponry, with results of all his weapons sorties. By using that information we will write an end-of-course report on him, which is broken down into phases, such as ground school, simulator, right through all the flying phases, ending with a summary by the Flight Commander and Squadron Commander, plus the Officer Commanding Flying, and finally the Group Captain who will add his literary pearls to it. The Flight Commander will be the one who has the most contact with the student, and he will run his course in effect, but the Squadron Commander will know him quite well too. The Wing Commander will probably have flown at least one or two sorties with the student, and what he is really doing is adding his experience to point out any particular areas which he feels are significant.

The net result is that we end up with three or four bits of paper that will act as a permanent record as to how he did here at the TWU at Chivenor. They will be sent to his next unit, his OCU, the Operational Conversion Unit, where the documents will be used as a guide by his new instructors, as they teach him to operate a front-line fighter. We try to be totally fair, as we're certainly not out to nail anyone. We try desperately hard to get them to succeed and pass them out. They have already cost the RAF a great deal of money, so the last thing we would want to do is to chop someone at this late stage. But on the other hand we have to remember that people's lives are at stake here, not only the student's own, but other crews', and maybe those of people on the ground, too. We don't go out of our way to create tension, and we wouldn't want every day to be like a driving test for the student. Every sortie is a test, in that they can pass or fail each one, but the aim of the game is to relax the guy as much as possible, as naturally we want to see him at his best. If we really wanted to, we could have a guy in tears before he even walked out to his aircraft, but obviously we wouldn't be achieving very much by doing that. At the end of the TWU course we will have a role dispersal meeting where we get together with the postings people and we work out a mutually acceptable agreement as to which student goes where. The postings guys will come along with the number of slots that are currently available, maybe four Tornado GR1 slots, two or three Tornado F3 slots, a couple of Phantoms and one Harrier for example. We might not have enough students to fill all the slots, but we will have a very good idea of which students will be best suited to each slot. Most guys do inevitably want to go to a single-seater, be it the Harrier or Jaguar, much of this being a natural degree of bravado I suspect, but they are all more than happy to go on to the Tornado or Buccaneer. Some will desperately want to go on to fighters, either Tornado or Phantom, and we have their measure of ability, the postings guys

have the slots, and so we endeavour to put round pegs into round holes where we can. Typically a student will have to wait something like six to eight weeks before joining his OCU, and after finishing here he will probably want some leave in any case, and precise times do vary. If we can't start him at an OCU within about ten weeks of completing his TWU course here, we will bring him back to do a refresher course of about five hours, just to get him back into the swing of things again, as we do want him to be as well equipped as possible when he arrives at his OCU. We don't want to send guys there who will not make it, because the further through training you go, the more expensive it is when someone fails. People do fail, and sadly we do get guys who are chopped from the squadrons, and when a guy has cost you over three million pounds you really don't want to chop him; but if the guy just cannot hack the job, or if safety is involved, you really have no other option. In many respects their hardest work will be done here at Chivenor, because whatever aircraft they may go to from here, it will have more advanced systems, things like inertial navigation … possibly not very reliable equipment in some cases, but something to hang their hat on. For instance, when we get people coming back here from the Tornado to join our staff, they have to work very hard to get back up to speed, because here they have no moving map in the cockpit to tell them where they are, or a navigator in the back seat whispering in their ear. Here they are back navigating by purely looking out of the window.

Aside from the actual flying training itself, life here is much the same as it was for them at Valley. We do expect them to act as big boys now, as they are out of Support Command, and they are expected to behave accordingly without any additional training on our part. They do perform Duty Officer activities here, looking after basic day-to-day matters, but most of the time here at Chivenor is really centred on the flying. They are almost encouraged to wear their flying kit most of the time, and we do try to develop a kind of social rapport between student and instructor. They are always told what is going on, and they are invited to squadron dinners and things like that. In every respect we just try to get them used to how things will be when they join a front-line squadron.

Today, the Hawk is still a very significant part of the RAF's training system, although the increasingly aged Hawk T1 is being replaced by the new Hawk T Mk 2, complete with new avionics and other improvements. Likewise, the T45 Goshawk remains at the core of the US Navy's training system and looks set to remain so for decades to come. Despite being designed as a trainer, the Hawk is a very fast and agile machine that possesses the capabilities of a fully fledged combat aircraft, and it is little wonder that the aircraft is operated by some export customers as both a fighter and ground-attack platform. It was the very last all-British military aircraft to be designed and built in Britain, and it is undoubtedly one of the very best.

4

VERTICAL VELOCITY

Many years ago, when a chance to fly in a Harrier was presented to me, I could scarcely believe my luck. There are few aeroplanes that enjoy such a unique place in history, but the Harrier is certainly one of them. It was the first combat aircraft to embrace the concept of vertical take-off and landing and it was of course a triumph of British technology and design, even though the basic principle of deflecting jet engine thrust (the very basis on which the Harrier was developed) was in fact a French concept. My initial excitement and eagerness at the prospect of flying in the magnificent beast was tempered somewhat by a long day of preparations at the Royal Navy's sprawling base at Yeovilton in Somerset, the Fleet Air Arm's 'home' and the shore base for the Navy's Harrier fleet. After first satisfying a doctor that I was capable of surviving the experience (something which brings a touch of sobriety to the joyous prospect of going flying), the flying clothing people then took charge, issuing the regulation thermal (and fire-retardant) underwear, flying overalls, gloves, boots, and the bizarre 'gsuit' that Harrier pilots routinely wear for every sortie. In fact it's hardly a suit at all. It resembles something akin to fetish wear, in the form of lace-festooned trousers which fit (with some difficulty) over one's thighs, calves and stomach, aided by long, tight zip fasteners. Once attached (and even this is quite a feat), the suit is laced up to fit snugly. Of course, the term 'snugly' is subjective, as once the gsuit is plugged into the Harrier's pneumatic system, the entire garment can (and will) inflate automatically when any gforce is applied to the aircraft, squeezing the wearer's body with a surprisingly firm grip. The aim is to restrict the flow of blood to one's extremities, thereby avoiding the risk of blackout.

The ubiquitous fast-jet flying helmet (the 'bone dome') is then fitted. The flying clothing folks wryly comment that the wearer's head is generally adjusted to fit the helmet rather than vice versa, and it does feel as if the screw fittings are reshaping one's head rather than the proportions of the helmet's inner core. But the bulky and absurdly heavy helmet serves a vital purpose, protecting the wearer not only from injury inside the aircraft but (most importantly) from injury should he be obliged to abandon the aircraft in flight. Leaving the Harrier in a hurry is a relatively simple task and something that can be achieved very quickly. It has to be this way as things can go wrong in a fast jet at any time, but in a Harrier things can go very wrong very quickly, often with no time even to think about the situation. Abandonment has to be instantaneous and instinctive, and the Martin-Baker ejection seat enables the pilot to make a rapid exit from the Harrier, simply by pulling the seat's operating handle attached to the seat pan. In most circumstances, regardless of height, speed or attitude, the seat will extract the occupant from the Harrier and place him under a parachute canopy in less time that it takes to read these words, capable of compensating for a descent rate of up to 100ft per second. But of course this ability comes at a price and the ejection seat is not some clumsy spring-powered prop from a James Bond movie. It's a tough, heavy, rocket-powered missile that makes no concessions for the comfort of its occupant, who must ensure that his limbs are clear of the aircraft structure (or they will be broken) and that he is properly positioned at the time of the ejection. Poor posture is not a matter of etiquette in the Harrier; it's the only way to avoid spinal injury. The clumsy 'bone dome' might seem unnecessary but should the ejection seat be used, its protection might be crucial. Attached to the helmet is an oxygen mask which (like all flying clothing) seems to be tailored to create as much discomfort as possible for the wearer. Hooked and adjusted with screw fittings, the mask incorporates a small flip switch that opens a microphone for communication. Another clip forces the mask firmly onto the wearer's face with a vice-like grip, making any attempt to speak quite an effort. This pseudo-torture is thankfully only necessary should the Harrier de-pressurise in flight or if fumes should enter the cockpit, the snug fit of the mask enabling oxygen to be supplied directly to the wearer. With the clear and tinted visors pulled down to touch the outer edges of the oxygen mask, one's face is entirely obscured by this bizarre arrangement of furniture. Intentionally so of course, as this ensures that the wearer's face is protected, particularly during the ejection sequence when the Harrier's canopy is shattered by an internal explosive cord. Fragments of Perspex and molten lead droplets are obviously not the kind of thing that one would like embedded in one's face.

Having survived the attentions of the medical team and the flying clothing specialists, I sit in a gloomy hall while a survival specialist explains the various support equipment that will be available to me, should I find myself on the ground or in the sea without my Harrier to protect me. The Harrier flight line still seems very distant. The inflatable dinghy, flares, food and distress beacon are common to most fast jets, but the Navy's safety briefings have a uniquely hands-on approach. Not only is a dinghy actually inflated for my information (and amusement) but I am also obliged to attach myself to a parachute harness before being suspended in mid-air, blindfolded. This might seem like a slightly grotesque party trick, but after having been spun for a few seconds whilst still blindfolded (to create even more disorientation), I'm asked to unfasten my parachute harness by touch, allowing me to tumble onto the (padded) floor. It's a good way to ensure that the survival theories have been understood and digested.

At long last the preparations are over and I am declared ready to go flying. The next morning I join No 899 Naval Air Squadron, the unit responsible for the training of new Sea Harrier pilots and the retraining of pilots returning from other duties. With an average of eight instructors and a similar number of students, 899 NAS was a relatively small unit, equipped with a handful of Sea Harriers and twin-seat T4N Harriers, together with a pair of Hunter T8M trainer aircraft. Students usually came to the squadron directly from the RAF's No 233 Operational Conversion Unit at Wittering, where they would have first spent thirty hours learning to fly the Harrier. At Yeovilton the students would then be taught to fly the 'navalised' Sea Harrier, using 899's mixed fleet of aircraft. The unit's Hunters were something of an oddity, being rather aged twin-seat examples of the classic aircraft, equipped with rather ugly nose-cone fairings that housed the Sea Harrier's Blue Fox radar. Fitting the radar into the dual-control Harrier T4 (which retains the early-production Harrier's thin nose contours) would have required major redesign work, and the Hunters provided a more cost-effective solution. No 899 NAS's twin-seat T4N Harriers were therefore used mostly for instrument rating check flights, initial conversion flying, and for teaching students the basics of launching from the famous 'ski jump' (commonly referred to on the squadron as 'the ramp') which equipped the Navy's 'through-deck cruiser' fleet. The T4N would also provide a suitable free seat from which I would be able to observe the Harrier's operation at first-hand. In order to show me what Harrier flying was all about, Charlie Cantan (a Falklands Sea Harrier veteran) was tasked with flying me as passenger on a combined training and demonstration sortie.

After briefing the planned sortie and getting myself kitted up again in my flying clothing, it is finally time to 'walk' and once Charlie has completed signing off our aircraft, he escorts me out onto the flight line at Yeovilton where a gaggle of assorted single and twin-seat Harriers are assembled and prepared for the day's flying schedule.

Although I have already spent many years as a photographer staring at countless Harriers, the aircraft takes on a rather different appearance when one is about to go flying in it. Up close, the Harrier isn't quite as sleek and slippery as it looks when it races past the crowds at an air show. In fact, the Harrier is quite a portly aircraft, essentially a pair of huge, gaping air intakes accompanied by a sharply downward-raked wing (itself covered by flat-plated vortex generators) and an oddly contoured fuselage featuring all manner of ugly bumps and protuberances. The Harrier is also surprisingly small and even at close range the aircraft gives an impression of compactness, particularly where its wings taper down to almost waist level. It's not often that you can stand on the ground and look down at an aircraft's wingtip. Even the twin-seat T4N seems quite small when compared to other combat aircraft, but the raised rear cockpit, longer fuselage and extended tail fin somehow make the 'twin-sticker' seem substantially larger than the diminutive Sea Harrier, even though they are essentially the same aircraft. The Navy's T4N was in effect identical to the RAF's T4 trainer, but without the add-on laser rangefinder nose fairing.

Our designated aircraft is already prepared for our arrival and while Charlie conducts a careful walk-round visual inspection of the aircraft (to ensure that no obvious faults have been overlooked and that locks and covers are all removed), I carefully clamber up the rather tall and shaky step ladder that is positioned beside the aircraft's rear cockpit. Small though the Harrier may be, climbing into the rear cockpit gives one the impression of being perilously high, but after stepping over onto the ejection seat (whilst holding onto the windscreen frame) I carefully route my boot-clad feet into the tiny floor space ahead of the seat pan, and gingerly sit down. The sensation of being perched high above the aircraft is now gone, replaced by an impression of clutter, metalwork, Perspex, dials, switches and pipes which festoon the tiny cockpit interior. My first task after sitting down is to lean forward and route the ejection seat's leg restraint cables through the buckles attached to my calves before clipping them into place on the seat's base. These cables serve no purpose during normal flight but if the seat's ejection sequence is initiated, they instantly draw themselves into the seat, pulling the occupant's legs firmly into place before the seat begins its rapid and violent departure from the aircraft. The main seat and parachute straps come next, and with the aid of the ever-helpful ground crew I am soon firmly attached to my seat and ready to attach the PEC which links the oxygen supply and radio link between the aircraft and its occupant, followed by the PSP (personal survival pack) clip which ensures that the dinghy and survival pack stay with you, should you need to eject. It also sets off a distress beacon that alerts the Search and Rescue force. Helmet on, radio link plugged in, oxygen mask clipped on, visor down, gloves on, straps tightened and everything is done apart from a few adjustments of the radio communications switches, to ensure that I can hear both the pilot and the air traffic broadcasts.

Once satisfied that everything looks 'good to go', Charlie climbs into the front cockpit and begins his careful setting up of the aircraft's switches prior to engine start. With the huge Perspex canopy pulled over my cockpit the surroundings become even more claustrophobic, and even though the rear seat is positioned higher than the pilot's position, the forward view in the Harrier isn't great, not least because there's a blast screen positioned between the two cockpits (in effect, a duplicate of the pilot's external windscreen) and the associated framework gives an impression of clutter and a typical 1960s feel which is also apparent from the cockpit's internal layout. No fancy CRT screens here, just a very traditional collection of dials and switches which look little different from those that were found in the Harrier's predecessor – the immortal Hunter. With a buzz of audio static and a crackle through my helmet's speakers, Charlie says hello from the front cockpit and informs me that if I'm ready, he can now start up the Pegasus engine. I am more than ready of course and a few seconds later a gentle vibration felt under my seat confirms that the engine is slowly spooling up, and even though my tight-fitting 'bone dome' is keeping out most of the noise, the distinctive whine of the Harrier's Pegasus is soon ringing through my ears. Unlike most jet engines, which can barely be distinguished audibly from each other, the Pegasus exhibits a noise which is unique to the Harrier, and even with the massive air intakes just inches away from my shoulders, the familiar Harrier 'wail' sounds much the same as it always has – just considerably louder. But with the canopies closed, the noise is greatly diminished and after confirming that my canopy is locked closed I remove the remaining safety pins to arm the canopy detonation cord and the ejection seat. Settled in the small (almost cramped) cockpit with a set of flickering instruments in front of me, my preparations are now complete and I can sit back and enjoy the ride.

ABOVE: ZB603 performs a rolling landing on a wet day at RAF Wittering. A cloud of water can be seen billowing upwards, created by the downward thrust of the Pegasus engine's nozzles. The markings of No 899 NAS are visible on the aircraft's tail. (*Tom Cheney*)

With the wheel chocks pulled away and a signal from the ground marshaller, the Harrier rolls forward, stopping smartly as Charlie stabs the brakes to check their effectiveness before turning left and off round the perimeter taxiway towards the runway. The journey is slow but comfortable, the Harrier's tough undercarriage demonstrating a smooth 'spongy' feel as the wheels ride over the taxiway's expansion joints. Ahead of us the Harrier's older cousins can be seen: a pair of Hunters setting off on their take-off run, embarking on another simulated attack mission over the English Channel. Once they are clear of the runway we are given permission to line up for our own take-off. It might be imagined that any Harrier flight might begin with a demonstration of the aircraft's unique vertical take-off abilities, but with a full fuel load (around 6,500lb) the aircraft is too heavy to be supported by the Pegasus engine's thrust and we are therefore required to use Yeovilton's very ample main runway to get airborne. A change in engine tone is the signal that we are about to go and as the brakes are released the aircraft lurches smartly forward with a rattle and roar of engine power, almost like a dog being released from a leash. Full power is applied, and this is the moment when the Harrier gives a passenger their first surprise. With some previous experience of flying in fast jets (including those with powerful, afterburning engines) I assume that the Harrier's non-afterburning engine will provide the aircraft with a fairly pedestrian take-off performance. Of course, I am forgetting that the Pegasus is a very powerful engine and that the Harrier is small and comparatively light. This combination of power-versus-weight is the key to the Harrier's unusual performance. Denied the opportunity to get us airborne vertically, the brutal power of the Pegasus engine is vectored away behind us, pushing the Harrier forwards at a truly breathtaking rate of acceleration. Rushing forward, we hurtle along the runway and in a matter of seconds Charlie is deflecting the engine's nozzles downwards and we immediately leap up off the runway into the air. Vertical it isn't, but still very impressive.

In a matter of seconds we are comfortably airborne, the landing gear tucked away (without so much as a perceptible clunk) and flaps up, and by the time that we swiftly skim over Yeovilton's perimeter road (where the hardy enthusiasts are busy pointing their cameras in our direction), there is little to indicate that we are flying in anything other than a pretty conventional jet. Indeed the initial climb-out seems indistinguishable from a similar experience I'd had at Yeovilton in a Hunter, which (by pure chance) had been one assigned to the Harrier Conversion Unit back in 1970, when the RAF first took delivery of the revolutionary aircraft. There is no doubt though that the rush of the Harrier's initial take-off acceleration is quite something when compared to the Hunter, but then with nearly twice as much thrust I suppose this shouldn't be a surprise. Climbing gently as we depart to the south-west, we quickly pop into a layer of milky grey gloom but just a matter of seconds later we are out over the top of the cloud cover in glorious sunshine. With little to do or see for the time being, I have the opportunity to examine the Harrier's interior more carefully, and with everything running, the hood closed and helmet on, the Harrier's 'office' is a cosy place. It is small by any standards and the instrument panel is correspondingly modest, with flight instruments to both sides of a large central display screen which dominates the panel. Above it, the HUD clutters up what is already a fairly restricted forward view (through an internal windscreen and then the forward cockpit), but looking to the left and right, the view of the outside world is commendably good. Panels on either side of the ejection seat contain yet more switchery, but the most obvious item here is the combined throttle and nozzle selector that sits by my left hand on the console. It's a small cockpit, but the elevated position above the front cockpit and nose tends to compensate for this, and high above Cornwall the view of the outside world is spectacular.

We are soon gently climbing through 20,000ft and ahead of us we have a rendezvous with a refuelling tanker. The tiny speck on the horizon eventually grows larger and as Charlie brings us in closer, we find that we are not the only 'customers' visiting the not-so-solitary VC10 tanker. Assembled in its wake is another VC10, two Buccaneers and our 'playmates' from Yeovilton in the shape of two Sea Harriers. Aerial refuelling was a regular part of Harrier operations for both the RAF and FAA, although the 'first generation' Harriers did not have the luxury of a more permanent (and retractable) refuelling probe. Instead, a large 'bolt-on' probe was attached to the aircraft when required, fixed above the port air intake and projecting forwards and outwards to the side of the canopy. It looked every inch an afterthought (which it was, as a refuelling probe was not built into the P.1127's original design) but it worked well and didn't affect the aircraft's handling in any way. The key to successful refuelling was for the pilot to concentrate on the tanker aircraft, taking visual cues from the tanker position (and alignment markings painted on its undersides), and to use only peripheral vision to nudge the Harrier's probe into the refuelling basket. Looking directly at the probe created a tendency to over-compensate and the result could be a slightly comical period of last-second frantic control column juggling which often ended in failure (and could result in a snapped probe nozzle or even a detached basket), but more experienced Harrier pilots had an ability to conduct the potentially hazardous task without any care for stipulated

ABOVE: The second AV-8B Harrier II, 161397, pictured here during developmental trials at Edwards AFB in California. (*Tim McLelland collection*)

procedures, and managed to make the art of 'tanking' appear to be far easier than it actually was.

With our time with the tanker completed, Charlie slowly edges our aircraft away from the VC10 and the attendant gaggle of customers, and radios the Air Traffic Control authority to request permission to descend. I assume that the descent will be as gentle and uneventful as our climb up to the tanker has been, but Charlie has other ideas, and although I don't know it, he is about to demonstrate what an agile and manoeuvrable aircraft the Harrier is. Charlie calls, 'Let's go down, are you ready?' over the intercom, I confirm that I'm happy and suddenly the world literally turns upside down. The Harrier flicks over with astonishing speed and as the horizon whirls round to a position above my head, we pull smartly downwards, the carpet of cloud below us suddenly filling the view ahead of me. We are heading down and heading down vertically like a brick, the altimeter 'clock' visibly spinning like something from a cartoon. In a matter of seconds we are slowly recovering to a more civilised attitude and as the cloud cover comes up to meet us, we re-enter the gloom and continue a more gentle descent, emerging above a patchwork-quilted Devon countryside, speckled with patches of sunlight as we head northwards and out of the miserable weather front. By the time that we've settled at lower altitude the cloud cover has almost gone and I'm able to enjoy a few minutes of sightseeing as we continue our journey back to Somerset in bright sunlight. But this is no time for joyriding, as Charlie wants to demonstrate what combat flying looks and feels like. As a Falklands veteran, he is no stranger to the art of low-level flying and as we turn to head north-east, he eases the Harrier's nose downwards slightly and gently descends through 2,000ft, then 1,000ft and still downwards until I notice that we're still edging downwards through 500ft. The gentle ride is just about over. The fields below are starting to grow significantly closer and give the impression that they are about to come up and wrap themselves around the cockpit, and as Charlie announces that we're now settled into the low-level leg of our sortie, the instruments confirm that we're just 250ft above the Somerset countryside, with trees, roads, farmhouses and streams flashing by either side of the cockpit, seemingly within touching distance. The sensation of speed is very obvious, but by looking directly forward

LEFT: The AV-8B incorporated many improvements over the original Harrier design. As can be seen, the aircraft features huge wing flaps, encased thrust nozzles, and modifications to create additional thrust-generated lift under the fuselage, between the external gun pods. (*US Navy*)

through the HUD, the world seems less disorientating. Looking out to the side at any passing object is almost pointless, as by the time that I've turned to look at it, we've gone past it in a dizzying 480-knot hurry. The key to gaining any spatial awareness is to keep looking forward. Things then get even more exciting, as Charlie adds another touch of power and begins a series of representative evasive manoeuvres, combined with an astonishing demonstration of knap-of-the-Earth flying. As the speed pushes through 500 knots, the gforces come on, squeezing me forcibly into the ejection seat, making my arms almost too heavy to move and my head feel as if it's made of lead. The gsuit inflates, squeezing my legs and guts, the horizon rolls left and right and as I've been briefed, I concentrate on squeezing my muscles to try and maintain consciousness while the gforce keeps

coming. I've been warned to expect this, but it's still a surprise when it starts to happen, and all of my attention is consumed by the simple act of staying awake and alert. Quite how Charlie manages to fly the aircraft is hard to imagine. Even more difficult to grasp is how the tactics of warfare can be entertained when the world is rolling left and right, the wings are streaming clouds of vapour and the cockpit seems to be a tight, claustrophobic, sweaty hell-hole. Still down in the proverbial weeds and still manoeuvring, we approach a looming hillside and as the forward view seems to suddenly fill with grass, the gforce comes on and we bang, rattle and thunder up the side of the hill, the gforce suddenly easing as the aircraft flicks over onto its back and we pull smartly over the brow of the hill. I look upwards through the top of the canopy and to my amazement a Land Rover flashes by, its occupant stood nearby staring straight up at us, seemingly just a few feet away. The gforce comes on again and – still inverted – we pull down the other side of the hill, snapping into a sharp roll which brings our wings level, and on comes the gforce once more as we

ABOVE: The RAF's Harrier GR5 cockpit was basically similar to the US Marine Corps AV-8B. Dominated by a Multi Function Display screen and a map display, the Harrier's unique throttle and nozzle level is visible on the left-hand console. (*British Aerospace*)

bottom out back in the weeds on the other side. Never was the term 'gob-smacking' more appropriate.

Charlie announces that he's finished with the manoeuvring (much to my relief) and we settle onto a direct heading for Yeovilton. Once back in the airfield circuit we nudge down for a low run across the airfield and a sprightly break to port, winding down the speed as the gforce comes back for the last time, until we're comfortably settled on the downwind leg of the circuit with wings level, speed knocked off and the undercarriage coming down. The steady road of airspeed is gone now, and the Harrier seems like a very different machine as we lumber around onto final approach, the engine tone now very evident. Crossing the runway threshold looks and feels entirely normal, but it's only now that Charlie demonstrates the uniqueness of Harrier flying. The airspeed continues to decay and as I look out and downwards, I can see that we're passing a speed at which any normal jet aircraft would have shuddered and stalled. But we're not stalling – the aircraft is slowing still further, down through 80 knots, 40 knots, until, amazingly, we seem to be at little more than walking speed, but we're still comfortably airborne, 50ft above the ground. And then it finally happens – the unique Harrier experience: we come to a complete stop. It's at this point that your senses defy your logic. The noise of engine thrust roars around you, even through the thick cockpit canopy and your sturdy helmet and ear cushioning. But we are stationary, hanging in mid-air, supported only by a column of pure power. The sensation is remarkable, and even more so because there is no feeling of 'artificiality' to detract from the very real feeling of jet thrust. The aircraft edges slightly forward, left and right, never entirely motionless, and the cockpit rattles with vibration, while the odour of hot kerosene drifts into the oxygen mask and up your nose. You really are left in no doubt that you are sitting on top of jet power. We start to move forward again, and the rudder pedals occasionally shudder, reminding me that the aircraft's air stream detection system is warning Charlie to keep the aircraft straight ahead, otherwise catastrophe would ensue and our flight would – at best – be ended courtesy of Martin-Baker's seats. But in Charlie's expert hands we're soon above the landing pad and as the grass begins to come even closer, we gently descend vertically until, with a last-second thud, we drop (almost bounce, in

ABOVE: An illustration of the Harrier's unique abilities: an AV-8B leaps skywards from a remote road during a USMC exercise. Although fully capable of performing both landings and take-offs vertically, most routine take-offs employed a rolling procedure, giving the aircraft some additional wing lift which was necessary if the aircraft was carrying a heavy fuel load or weapons. (US Navy)

fact; the undercarriage is surprisingly springy) onto the concrete pad as the roar of the Pegasus simultaneously winds down. Back on the flight line and with the canopy swung open, I'm able to gather my thoughts and carefully clamber back onto the concrete, soaked in sweat, tired, my head still spinning. The sheer rush of a fast-jet mission is enough to thrill anyone, but to end the flight in such a remarkable and unique fashion is something that can – or at least could – only be achieved courtesy of the Harrier. By any standards, it has been quite a ride.

Despite being a uniquely British design, it was with some irony that Britain was the first Harrier operator to dispose of its assets. Thanks to a rapidly contracting defence budget, the RAF and FAA abandoned their Harrier force in 2010 and most of the RAF Harrier GR7 fleet eventually found its way across the Atlantic (by ship) for use as a spares source for the US Marine Corps Harrier fleet. Unlike Britain, America saw no logic in abandoning its Harriers and they will remain in front-line service until the new F-35 becomes established. Captain Nicholas Dimitruk was one of the USMC's elite band of Harrier pilots and in response to a barrage of questions from myself, he ably described first-hand the world of operational Harrier flying as seen from the pilot's perspective:

The Harrier serves many different roles within the United States Marine Corps, but it is primarily used as an air-to-surface weapons platform. Most would say that within the air-to-surface realm, the Harrier focuses on the role of CAS. Close Air Support, by definition, is the exertion of air action by fixed or rotary winged aircraft against hostile targets that are positioned close to friendly forces, requiring detailed integration of each air mission with fire and the movement of these forces. Other missions that the Harrier can support include Air Interdiction, Armed Reconnaissance, Strike Co-ordination and Reconnaissance, Air-to-Air, Anti-Air warfare and Forward Air Controller (Airborne). Any Harrier mission can involve any combination of these missions, but as noted before, Close Air Support is the focus, and the most prevalent mission in today's warfare. As a pilot in a Harrier squadron, you will generally know what type of mission you will be flying as much as a week in advance, but of course you can be scheduled on the spot because you are always expected to be able to execute any type of mission that you are qualified to perform. Generally, but depending on the complexity of the mission, planning for a mission begins several hours prior to the brief time. The Harrier has two computer screens called Multi Purpose Control Displays that sit side by side and provide several different functions for different portions of the flight, but most notably they contain a moving map and compass rose, both of which are key to giving the pilot complete situational awareness. During the planning stage, most of the map that the pilot will see

LEFT: AV-8B Harrier II Plus from VMA-542 pictured in the hover just seconds before touchdown. Devoid of any external stores, the aircraft performs well in the hover and carrier recoveries were consequently simple to achieve. Take-offs (with fuel and weapons loads) were made in 'rolling' mode, achieving wing lift to combine with engine thrust. (*US Navy*)

in the flight is first created on computers in the planning room. You have the ability to select accurate waypoints, build routes, create map overlays, satellite imagery, calculate times, build weapon envelopes and so on. The computers can also help calculate weights and drag, plug in frequencies into preset channels, and 'weaponeer' diagrams help give the pilot a visual representation of how an attack will be conducted. Almost all aspects of the mission can be pre-planned and plugged into the computer, which can then be transferred to the jet. We always like to have paper copies to back up the plan should the computer fail, so we will often have waypoint locations, radio frequencies, weaponeering diagrams, weights and drag and so on, that can be manually fed into the jet if necessary. This of course is the backup plan and can be very time-consuming, and as any pilot

knows well – time equals fuel. Once we build our missions and save them to a disk, we will usually get another weather update and check the notices to airmen (NOTAMS). After several hours of pre-flight planning, we are then ready to conduct the brief.

Briefs always begin exactly on time, and the first item of business is to conduct a 'time hack'. We work backwards from when we need to have the bombs on target. This allows us to know when to take off, when to taxi, when to check in, when to start the jets and so on. These time hacks are essential for us to manage timelines in such a manner. Following the time hack, we conduct a product inventory to ensure every participating pilot has the proper products to aid him in the flight. This includes maps, imagery, and what we like to call 'smart packs' that include all necessary information pertaining to the flight. Following this, we conduct a sortie overview and outline our main objectives. This allows everyone to see the big picture before we start getting down to the exact details of each part of the mission. Intelligence will brief us on enemy and friendly dispositions, allowable risk in regards to threats we might face, threat specific details such as the envelopes in which the enemy can engage us, or what indications our cockpit will give us if we were to be engaged by a surface-to-air missile, etc. Following an intelligence report, we tackle the administrative portion of the flight. This includes current and forecasted weather, call signs, aircraft assigned, block altitudes assigned to each aircraft for de-confliction, the succession of leadership should the lead jet break down, what ordnance our aircraft will have, how much fuel, how heavy we expect to be, what our drag will be, and what we can expect in regards to backup jets should our jets break down. Next we will go over our VSTOL capabilities. The ability to hover, or any nozzle usage other than when directed aft, relies on our aircraft weight, expected weather, and elevation (altitude). For weather, the colder the temperatures and the higher the altimeter setting, the more thrust our engine will produce and the more likely we will be able to hover. The same applies for elevation – lower elevations allow greater engine performance. Added to this, the aircraft's gross weight has to be sufficiently low enough to hover. For the pilot, hover weight is generally achieved by burning gas down to an acceptable level and hopefully returning without any heavy ordnance. In the jet, our computers calculate how heavy the jet is and provide us our hovering capabilities in the form of fuel levels. In other words, it will tell us at what fuel level we need to be at before we can start hovering with the current aircraft configuration. If it spits out 4.5, for example, then I will need to burn my gas down to 4,500lb before I can expect to be light enough to hover. If it gives me a number close to zero, or a negative number, then I am being told that under the current aircraft configuration and weather, my aircraft will not be able to hover. So during the VSTOL part of the brief, we like to get an idea of whether we will have VSTOL capabilities, and if we don't, what methods we can use (such as jettisoning bombs) to reach hover performance in case we need to do so in an emergency. If we are away from home

base, we usually reserve this portion of the brief to also go over other such variables that may affect both our departure and arrival into a foreign field. Such variables include field elevation, runway distances available, what speeds we can reach on the runway before we are no longer able to abort the take-off, whether there is concrete available should we have to perform a vertical landing (our exhaust is so hot that asphalt will melt, thus requiring concrete for vertical landings) and so on. We next cover fuel management. We usually designate three fuel levels that we can program into our jet. Upon reaching that fuel level, the jet will give us both an audio and visual alert. We begin with 'Tiger' fuel, which we usually designate as the level of fuel at which we can afford one more attack or maneuver specific to the mission. Next we have 'Joker' fuel, which usually designates enough fuel to stop the tactical portion of the flight, join up, and go home. The last fuel state is 'Bingo' which is the fuel level that means we need to be pointed at home base immediately. We will also use this portion of the brief to discuss details concerning the refueling tanker should we be requiring in-flight refueling. For instance we'll discuss the tanker times, what sequence different sections are going to rendezvous the tanker, and any other fuel-related administrative briefing items.

Next we cover communications. We will cover what frequencies we are likely to use, which preset channels (1 through to 99) that we have assigned each UHF frequency, what frequency we will use for inter-flight communication and what frequency we will hop to should we have no luck on that frequency (referred to as 'chatter mark' frequency). We should normally have two working radios, one for inter-flight communications and the other used primarily for external agencies. Sometimes one of the radios might not work, so we have to cover how we will handle this particular contingency. We will also cover the plan should we lose communications altogether. Next we cover navigation and we usually cover our waypoint plan at this stage. We use waypoints to mark our home field as well as alternate airfields, points of interest along the route, locations where we would like to filter through to both begin and end an attack, and of course the targets themselves. We will usually print out maps of several scales to give us better situational awareness of the where the waypoints are located. We also like to use satellite imagery to find good geographical reference points that serve to help talk each other onto targets visually. We will discuss our sensors and how effective they will be given the time of day and the environmental conditions. We have many sensors that can be used both by day and night, including a forward-looking infrared pod, radar, camera, and night vision goggles. We have two FLIR (Forward Looking Infra Red) systems, one of which is internal to the jet and used primarily for navigation. The other is located in a targeting pod that is hung off our wing. As the name suggests, this FLIR is used primarily for reconnaissance and targeting. Unlike the FLIR internal to the jet, this FLIR can be slewed about its axis. Environmental conditions such as visibility, humidity, time of day, etc., greatly affect how well these sensors can operate. For instance, the position of the sun greatly affects our ability to see targets, especially when it comes to shadows. Similarly the position of the moon greatly affects our ability to see at night. We will have dedicated weather personnel send us over a report that details how well our sensors are expected to work that day and how far we can expect to see given the variable conditions. We will then go over our weapons loads and how we need to set them up and employ them so we maximize our ability to accomplish the mission. This is a very detailed process with many moving parts, and any kink in the chain could result in non-functioning bombs or unsafe attacks. We will go over how to pre-flight each weapon, the limitations of the weapons themselves, and the limitations that these weapons place on the jet. Each weapon and external store configuration limits how fast we can go, how fast we can release the weapons, what G's we can pull, how steep we can dive, whether we can dive at all, and so on. It's also important to cover how many weapons we can release at any one time and the time intervals between each release should we choose to drop multiple weapons. We calculate the distance between where each weapon will impact the ground, and what will maximize the probability that the specific target will be destroyed. Discussion will cover not only how we will deliver the ordnance but also how we will fuze it. How long, from the time the weapons falls off our jet, do we want until the fuze arms the bomb? Do we want the explosion to happen upon impact, or do

we want there to be a delay so that the weapons can penetrate a roof for example? As with most aspects of flying, we always like to plan for contingencies and worse case scenarios. What happens if the bombs fail to come off the jet? Will we be too heavy to vertically land? Will we be too heavy to land at all? What happens if only the bombs on one wing come off, leaving us with an asymmetric load? Will we have enough aileron control authority to keep the jet controlled? Will the puffer ducts on the wingtips have enough thrust to keep the jet upright should we need to use our VSTOL capabilities and we happen to be asymmetric? What do we do should our gun misfire, or we have hot round in the chamber? As you can see, you can never plan enough, especially with the endless possibilities that come with the task of delivering ordnance.

Much like we can input fuel levels to alert the pilot, we can also input altitudes to give us more situational awareness. For each type of delivery, we can input an altitude that will flash a warning light and audio tone to alert us when we pass through that altitude. When we dive deliver a bomb, we will usually use this feature to mark the minimum altitude in which we can make a safe recovery. In other words, when we hear what we refer to as 'bitching betty' saying 'altitude, altitude', it means we need to start pulling up now before we hit the ground or 'frag' ourselves (fragmentation damage) with our own bomb. This can be a real lifesaver, especially at night when we are diving into pitch black. We can also input an altitude into the computer that gives us a visual cue in our head up display to show whether the bomb will dud or not. As we dive towards the ground we lose altitude, which means when the bomb leaves our aircraft it has less and less time for the fuze to arm the bomb. Eventually, at a given altitude, there will not be enough altitude for the fuze to arm the bomb, and as a result the bomb will fall harmlessly and dud. So we will discuss what altitude we need to input into the system, given our calculations, so that we get an accurate visual cue in the HUD that notifies us when our bomb will dud. Finally, given all these variables, we will go over the numbers of each type of attack to include altitudes, speeds, dive angles, etc. The Harrier is armed with the ALR67, a passive detection system designed for both airborne and ground threats, and we will set up this system. Do we want the computer to prioritize air threats or ground threats? Do we want to display all threats or do we want to prioritize only those threats most dangerous to us? We are also armed with an ALE39 that is a system that dispenses flares, chaff, and jammers (expendables). We use the heat from flares as both a pre-emptive and a reactive way to decoy infrared guided missiles. For both radar and radar guided missiles, we use chaff as a decoy. Finally, we can expend jammers that will jam specific radars and radar-guided missiles. During the brief, we need to know what and how many expendables we have and how we plan to set up them up. When we activate the flair switch for instance, how many flares do we want to go off, what time interval between each flare do we want, when should we activate this switch both pre-emptively and reactively, etc., and then we will discuss the same in regards to chaff and jammers. After these systems discussions, we then talk about how we plan to taxi, what type of take-off we want, the departure, arrival and landing that we plan to do. What formations do we plan to be in during each portion of the flight? How fast, and at what altitude do we want to be? What approaches at each airfield are available to us should the weather go bad. We will then discuss what external assets are available to us for the mission. Will we have tankers for in-flight refueling, will we have EA6B Prowlers to jam enemy radar, will we have AWACS to provide us de-confliction and be our eyes in the sky, will we have a forward air controller on the ground helping us guide our bombs to their intended target, and so on. Lastly, we will go over all the possible emergencies we could encounter from weapons malfunctions to bird strikes, engine fires, oil failure, mid-air collisions, ejection procedures, etc. We will also discuss what our game plan will be for each emergency. We will explore what particular rules we have to help mitigate any of these emergencies. Every particular type of mission, whether it is close air support, low-altitude training, air-to-air combat, air-to-surface or anything else, has a set of rules that we legally have to verbalise before we can fly that type of flight. This covers the administrative and tactical administrative portion of the brief. Everything discussed above should typically take fifteen minutes of the brief. As you can see, it is a lot of information to be passed in such a small period of time. This is why, as I stated before, flight leads are trained to be extremely articulate and well spoken. The majority of the brief is left to the tactical portion of the brief. This tactical portion will discuss in minute detail every maneuver we plan to do during the actual attack phase of the flight.

The many different types of attack we can do would be difficult to summarize succinctly. The tactical portion of the brief usually concentrates on specific attacks we plan or are likely to conduct during the flight. The flight lead will first cover the goals of the mission that determine mission success. The flight lead will go into minute detail covering every second of the attack timeline. He will discuss every altitude, position, heading and airspeed he expects every member of the flight to be in throughout the entire course of the attack or maneuver. He will discuss every check each pilot needs to do, what the pilot should see in his head up display, and what he should see in his computer screens below. He will discuss common mistakes and how to correct these mistakes. He will go over required communication, expendable usage, and all details pertaining to the attack. The flight lead will have every attack drawn up on the board and will likely use plane models as a visual aid. We try to be as standardized as possible, which means every word uttered by the lead must be in accordance with what the tactical manual says. Following the brief we get one last out-brief from the duty officer. This duty officer (always a pilot) will provide a final report on weather and any significant details concerning the airfields we plan to use. He will make sure we have jets assigned to us and that they are ready to go. He will make sure we understand all the risks pertaining to the flight and that we have read all current squadron policies. Lastly he will run down a quick checklist to make sure we have accomplished all basic items required for the flight. The next order of business is to walk to maintenance control and sign the jet out. Every jet has a history of what maintenance it has received recently, and it is essential that each pilot has an understanding of what unique characteristics and discrepancies are associated with that aircraft. This is also an opportunity for the pilot to raise any questions before he finally signs for and takes full responsibility for that aircraft. We will also be able to confirm at this stage that our aircraft has the proper ordnance and external stores loaded. Following this we will suit up.

Our equipment includes a GSuit, harness, survival vest, and helmet (with or without NVGs). As we suit up we are performing a quick inspection of the equipment. Are there any holes in the GSuit? Is the harness twisted? Do we have all required survival items in the vest? If we were to use NVGs, we would use this portion of the flight to adjust and focus them. The Flight Equipment room has a special dark box with an eye chart that allows us to properly focus the NVGs.

It is then time to walk to the jets and conduct a pre-flight inspection. The aircraft at this point has been heavily prepped by the plane captains and maintenance personnel, and the pilot's pre-flight is just one last check to make sure the jet is ready to be taken into the air. The pre-flight is designed to be one single walk around the jet, checking that all external probes look like they are in functioning order, that panels are secure, no bolts are missing, control surfaces and puffer ducts look to be in working condition, intake and exhaust ducts are clear of any foreign objects such as tools and debris, the aircraft has sufficient oil levels, the tires are in good shape and so on. We will also do a quick foreign object sweep in front of the aircraft and then climb the access ladder to begin strapping in. Before we climb in the cockpit, we check the seat to make sure all safety pins have been removed (so that the seat can eject), make sure all the straps are good, switches are in the correct initial position, tapes and cards (that hold the mission plan much like a CD) have been inserted and are ready to go. We climb in, strap in, and prepare the cockpit for startup. In tactical aviation we test our systems using what we call BITS (Built In Tests System). With the flick of a switch we can perform BITS that will help to show whether the system is working as it should. If the system is not performing properly, then we will be given a certain indication as to what is wrong with it. We perform a BITS check on almost all of our systems. Once the cockpit is prepped for startup, we check our watches and start the engines at the precise briefed time. We flick the engine switch to start, watch as the RPMs (Revolutions Per Minute) creep up, and with sufficient initial RPMs, we push the throttle up a little to introduce fuel into the combustion chamber. The RPMs continue to creep up along with the engine temperature, and we are looking for the engine to not only light off in time but to reach a specific RPM at or below a specific temperature.

If we see any abnormality in the start that exceeds a specific value (for instance, the temperature of the engine during start exceeding 475 Celsius) then we will immediately cease the start process and try to troubleshoot the cause of the problem. With a good start, we can continue the process of getting the jet ready to fly. We test the hydraulic system, make sure all the lights work properly (such as warning and caution lights, or gear down/unsafe lights), we check the engine numbers, load the mission planned data into the plane's computer, turn on the aircraft's navigation/communication and other electrical systems, perform tests on the fuel system, the electrical system, the water system, check the trims, the standby instruments (which are the traditional stem gauge instruments provided for if our electrical ones should fail), the oxygen system, and the flaps system. We cycle all of our controls to make sure they move properly and that there is nothing binding them, and we conduct a mass BITS check that pretty much tests all other systems to make sure they are functioning properly. After making sure the jet is ready to fly, we then prepare it to make sure it is ready for combat. We check that we have proper GPS time loaded, we make sure the aircraft computer knows what weapons it has loaded, what fuze times we want for each weapon, what system we want to utilise to drop the weapon etc. We will double check that our GPS is operating at a sufficient level, that the targeting pod is properly calibrated, and that our expendables are properly loaded (chaff, flares, jammers). We check within each of us in the flight that all our radios work properly. Once the jets look to be in good shape and all weapons are set up sufficiently and armed by the ordnance personnel, the lead aircraft will communicate with ground that we are ready to taxi, and we are then ready to leave the flight line. We will almost always taxi sequentially and with a considerable distance between each other ('considerable' meaning more than in other tactical jet aircraft). As mentioned before, we are very wary of foreign objects (such as rocks and debris) and the risk of such items entering our intakes and damaging the engine. As you will know, the Harrier has very large intakes, so more than most aircraft it is very susceptible to foreign object damage. We taxi very far apart to prevent another aircraft's exhaust from blowing debris down the trailing aircraft's intake. As we roll towards the runway, we will conduct our final checks to prep the aircraft for take-off. Part of this means checking (given the current weather conditions) what speed we want to throw our nozzles down to jump off the deck successfully and what angle we want the nozzles to be at to maximize our lift (we have a stop near the nozzle lever that can be moved to a particular nozzle angle so we can quickly throw the nozzles to this position). We set our trim to the proper settings, check our flap setting, and double check we have no warning/caution lights.

When we are ready to take the runway, we notify lead that we are 'up and ready' and he will call the tower to take position on the runway. We position ourselves on the runway usually with the same spacing as before (for FOD reasons), and the wingman notifies lead that he is 'one finger checks complete'. This is terminology to signify that he is ready to run his engine up. Lead calls for take-off clearance and tells the flight to run up the engines. This is our final check before we go airborne. We run the engines up to 60 per cent and check that the engine accelerates properly and that all the moving surfaces internal to the engine move to the right position (we have an engine page on our computer displays that provides all of this engine data), we throw our nozzles to the position we want for lift off to check the nozzles and nozzle indicators work, that the flaps program properly alongside the nozzle movement, and that engine duct pressure going to our puffer ducts is sufficient. We throw the nozzles back forward and tell our lead we are 'two finger' which means final checks are complete, and that my jet is ready to be airborne. Lead will roger up this call and say he's rolling. Lead conducts his take-off and as soon as the lead aircraft breaks the deck, the wingman is then ready for his take-off. For take-off, we hold a button to command proper take-off steering, we release the brakes, and slowly and smoothly add full power. Initially the engine will spin up relatively slowly, but as the engine speeds up it will accelerate very suddenly. The pilot will experience a very sudden acceleration which, when experienced for the first time, is an amazing rush. Not many (if any) aircraft can replicate the Harrier's ability to accelerate below 10,000 feet. As the engine reaches full speed, we check that the engine is operating within its RPM and temperature limit. If there is a problem, it is most likely too late by now to abort. Because the engine accelerates the

aircraft at such a rapid rate, the aircraft will be too fast to successfully abort. It is usually more appropriate to take the aircraft into the air at this point and troubleshoot the problem airborne, rather than trying to control and brake the aircraft within the remaining runway length without blowing the tires and losing control. That being said, we would have to abort should the emergency prevent us from flying (such as a complete hydraulic failure). With the engine spooled up and the aircraft accelerating down the runway, we very quickly hit the velocity (calculated before by our flight computer) that we take our nozzles from fully aft to the specific angle calculated before. As the nozzles are thrown down, the aircraft literally jumps off the deck like a bottle rocket, aided by some of the aerodynamic lift from the speed that we already have under our wings. The next process is a gradual trade-off of vertical force between the nozzles blowing exhaust downwards and the lift created by our wings. As we creep the nozzles back to their aft position, less thrust is vectored downward so we lose this lifting force. On the other hand, as we push the nozzles to the rear, more force is vectored horizontally accelerating our aircraft and causing the wings to produce more lift. So as we are losing the jet-borne vertical component of force, we are gaining more lift through the wings. As stated before, this is a careful trade-off to keep the aircraft airborne and in control, and a trade-off that is 100 per cent under the control of the pilot. If done too quickly, the lift from the wings will not be sufficient to compensate for the loss of downward thrust from the nozzles, and the aircraft will descend and possibly hit the deck. If done too slowly, the aircraft will remain in a limbo of jet-borne and wing-borne flight that can overheat the engine and also cause the aircraft to linger at a low and dangerous altitude.

As we perform this slow and controlled acceleration, the pilot will lift the gear up before reaching 250kts, re-program the flaps for conventional flight, and eventually we will be flying like any other conventional aircraft (nozzles fully aft and all controllability through our aerodynamic control surfaces). We run down our lead aircraft and pull up into formation. As we fly out to the tactical area, we start setting up our systems so we are ready to attack the given target. This involves arming up the system, changing our head up display image to the air-to-surface mode (or air-to-air mode given the mission at hand), selecting weapons and checking the right settings, un-stowing our targeting pod and checking that the display is clear and properly focused, and then arm our flares/chaff and jammers. We access the environmental conditions such as the visibility and the wind condition at altitude. The wingman will pull up next to lead and both aircraft will expend a single flare/chaff and jammer to verify the system is working properly. Next the aircraft will perform a couple of high G turns (called a Gwarm) to make sure that the pilots are in a proper physical condition to take Gs and that the aircraft itself is in good condition to pull Gs. Next we will test the laser located in the targeting pod. The laser has many uses. It is used to calculate very accurate target elevation (one of the most important and difficult variables when trying to generate the precise location of a target), it helps guide bombs to the target (i.e., a laser detector located in the nose of certain bombs hones towards the target that the laser is pointed at) and it also acts as a laser pointer to help steer your eyes towards the correct target. To test the laser, one aircraft will fire the laser while the other aircraft's laser detector will pick up the energy and point the targeting pod to where the laser energy is coming from. When both pilots confirm that the laser works and it is pointing to the desired location, the planes will switch roles to test the other aircraft's laser. Following this, we perform final checks on the targeting pod, check our lighting package (securing external visible lights especially in combat) and double check the weapons and expendables. At this point we will notify our flight lead that the aircraft is completely set up and ready for the tactical portion of the flight.

During the outbound leg to our assigned area of operations, we will go through several controlling agencies. We pass them information about our flight (such as the ordnance we are carrying) and they filter us down eventually to the unit they want us to support. As stated before, it is usually known in advance which area or unit we have been assigned, but we are always ready to be on call for whoever needs us. Eventually, for a Close Air Support Mission, we will be in a holding stack above our assigned area and be in contact with our eyes on the ground (in the USMC this will usually be a Forward Air Controller – another 'Winged Aviator'). After checking in with the

FAC and providing him with pertinent information concerning the flight, he will give us a quick update on the situation on the ground. This usually consists of both the enemy and friendly location and situation. Following this, and should the Marines on the ground need it, we will begin setting up an attack. The Marines, as well as all of the other services, have a very standardized way of setting up a CAS attack. The FAC on the ground will pass all key details of the attack in what is referred to as a 'nine line'. The nine line is essentially nine pieces of information needed to conduct the attack. The Harrier has a CAS page in its computer devoted to this nine line. The nine line consists of the following – the location to begin the attack, the bearing and distance from this location to the target, the elevation of the target, the description of the target, the exact coordinates of the target, how the FAC will possibly mark the target (laser/smoke), exact location of the 'friendlies' and finally the route we need to egress the target area. Following this, he will give us the time he wants the ordnance to hit and the heading he wants us to attack from. Following the nine line, we must legally read back the target elevation, location, the final attack heading and any restrictions he gave us. Once we are confident our system has what the FAC has passed to us, we input the information and load it into our maps. The digital maps will automatically provide a display showing the location to begin the attack, the target and egress route, the attack heading, and the location of the friendlies. All the aircraft in the flight will make sure everybody is working in unison ('looking at the same sheet of music'). The lead will describe where he sees the target in relation to the digital map, the bearing and distance to the target, the speed required to hit the target on time, what attack and what weapons he proposes to use, what time the flight will actually push to the target, and finally the flight will physically plot the target on a map and describe what they see. This entire process is to make sure everyone is on that same proverbial sheet of music. Once the flight is confident of this, the aircraft can snap the targeting pods so that each pilot can see the target given by the FAC on their computers. At this point the FAC will help talk us on to the specific target he needs to be attacked. Though coordinates can be very accurate, we always make sure that we have the right target and a FAC 'talk on' is just another way to ensure this. Once we are absolutely sure we have the correct target, the flight will push to the initial position from where the attack will commence.

The Harrier has many different types of attack catered to both the types of situation and the type of ordnance carried. For laser-guided weapons, we have the capability as a single aircraft to both drop and laser a weapon to the target. If one of the lasers from the targeting pod is not functioning properly, we can have a wingman laser a bomb dropped from the other aircraft in the section. The laser designation can also come from another platform such as an attack helicopter or even the FAC on the ground. If there is a cloud deck, we also have the option of having an aircraft above the clouds drop on a coordinate that will put the bomb in what is referred to as the laser basket, and have another aircraft below the cloud deck with eyes on the target laser the bomb to the exact target. The other type of precision bomb we have is the Joint Direct Attack Munition (JDAM), which is essentially a GPS guided bomb. With our targeting pod, we can laser and generate very accurate coordinates that can be directly transferred to the JDAM. Once we fly into the weapons envelope, we can release the bomb and it will guide itself accurately to the target. Both the JDAM and the Laser Guided bombs are usually dropped from straight and level flight. For general munitions (dumb bombs as we like to call them), we will usually dive deliver. Though the bomb is literally just a free-falling piece of explosive, we can still utilise our targeting pod to enhance their accuracy. The mechanics of finding a target and lasing it for precise coordinates are the same as that for the precision guided munitions. These precise coordinates are fed into our head up displays so that we get an accurate visual representation and cue as to the exact time and position to drop to the bomb. Though less accurate than the LGB and JDAM, it still provides very accurate delivery. We can also use cluster bombs and firebombs which are delivered very similarly to the dumb bombs. For forward firing ordnance, we are armed with a 25mm Gatling gun and 2.75 to 5 inch rockets. This type of ordnance is delivered in a simple ramp down attack with the nose pointed at the target and firing at a specified altitude. We also have the LMAV, a laser-guided forward firing missile for surface targets (very useful for moving targets).

For air targets, we can carry an assortment of air-to-air missiles to include the AIM9 (IR missile) and the AMRAM (radar-guided missile).

As we push from the initial position to the target, lead notifies the FAC that the flight is inbound. The flight begins a set of maneuvers designed to accomplish the lead aircraft's scheduled moves. For instance, should the lead aircraft want two bombs dropped with a minute interval, the wingman needs to fly a separation maneuver to make this happen (using geometry and timed turns). Along with these maneuvers, both aircraft are busy manipulating their targeting pods so that the most accurate target coordinates are being fed to the system each second. As each aircraft completes all the final checks to make sure the system is set up and as the aircraft nears the point of ordnance delivery, both aircraft will confirm they are visual with the target and their laser is firing. Whether it be a straight and level or dive delivery, each aircraft will call 'in' with a heading. This call tells the FAC that the aircraft is ready to employ ordnance from the correct heading given in the nine line. Ordnance delivery all rests on the FAC giving either the 'Cleared Hot' or the 'Abort' Call. With a cleared hot, the pilot's thumb presses the firing mechanism for the specified weapon. The other aircraft will always have his targeting pod in the area to assess the hit of the other aircraft. Once the aircraft have employed their ordnance, they will meet up at either a higher 'sanctuary' altitude or just 'low and away' from the target and either return home or set up another attack. Following the tactical portion of the flight and during the return to base, lead will call the flight to 'fence out'. This command requires both aircraft to secure their Master Arm switch, switch from their 'air-to-ground' tactical display in the HUD to the navigation mode, de-select any weapons, secure the targeting pod, and turn the expendables off. Lead will then clear the wingman into a close formation and, provided all his weapons are secure, a battle damage assessment will be conducted. Both aircraft will take turns inspecting each other for any damage that might have occurred during the mission. Most likely you are looking for loose or missing panels, possible bird strikes, and possibly in-combat damage inflicted by the enemy. While this occurs, lead is performing and delegating administrative tasks such as passing on mission reports and intelligence to the airspace controller, giving a battlefield turnover to the flight that takes over, checking up on the weather at the destination, and checking the fuel level required for aircraft to possibly land vertically etc.

Before or after the tactical portion of the flight, the Harrier has the capability to refuel whilst airborne. The tanker usually has a pre-planned refueling track, and the flight approaches this tanker track from below to ensure altitude de-confliction. After locking up the tanker on the radar, and having the tanker clear us aboard, the flight joins the tanker from below and to the left. Each flight member prepares the cockpit for refueling. The Harrier, like most tactical jets, does not favor slow flight. Unfortunately, because many tankers fly in the low 'two hundreds' for airspeed, we are forced to fly the same airspeed if we want to refuel. To help the jet fly comfortably at this speed, we will usually choose a flap configuration (primarily used for landing) to maximize our lift at this slow speed. We extend our fuel probe and check the aircraft is flying in co-ordinated flight (the fuel probe can significantly change the aerodynamics of the Harrier especially in yaw, so significant rudder trim is required to accomplish co-ordinated flight). Once the aircraft is ready to accept fuel, the tanker will clear us in. Unlike many aircraft, the Harrier has a fuel probe that extends from the left of the aircraft (rather than emerging straight out in front). As a result, it takes some practice to steer the probe into the basket. We often approach the basket looking straight ahead and use our peripheral vision to place the probe near the basket. If we were to stare at the probe, we will likely create pilot-induced oscillations. This is a phenomenon whereby as the basket naturally sways to and fro in some direction, the pilot over-corrects to keep the probe aimed towards the basket. These over-corrections usually result in a counter-oscillation making successful contact with the basket near impossible. Our peripheral vision is less susceptible to this, so with the unique location of our probe, this is what we use to initially guide the probe to the basket. As we approach the basket from the side and it is no longer visible in our periphery, we take a quick glance at the probe and make a last-minute correction if needs be, to put it in the basket. With successful contact, we push the hose in to initiate fuel flow to the aircraft. Once we have filled up, we tell the tanker we are satisfied and he clears us to break away (we then move aft until the probe un-sticks from the basket) and form the flight high and to the right. Following everyone's successful refueling, we are cleared to break away and conduct the next portion of the flight. Once both aircraft have completed all of the administrative tasks, and the flight nears its destination (a ship in this case), both pilots ready their cockpit for landing.

The typical landing at the ship will have both aircraft flying a tight formation at 350kts overhead the ship at eight hundred feet. As the flight continues a couple miles past the ship, the lead will give his wingman the signal that he is going to break, and will usually give the signal for the break interval, which is generally two seconds. As lead banks his aircraft sharply and turns, wingman counts a couple of seconds and does the same. We bank our aircraft to near 90 degrees, pull our throttle to idle, extend the speed brakes (a switch on the throttle will open up a flap door on the base of our aircraft to create drag and help slow down the aircraft), and pull approximately 4gs. The aircraft will slow down quickly, and as we complete 180 degrees of turn we can generally expect the aircraft to be below 250kts (the max. speed that we can pull the gear down). We throw down the gear handle, switch the HUD to the VSTOL display (a display that provides us all the critical information to safely perform a landing), switch our altitude display from barometric to radar (which gives us height above the actual ground rather than sea level), and at this point we are ready to utilise the Harrier's unique VSTOL capabilities. As we ease the nozzles from the aft position, we switch to a flap program that allows us to maximize the ever-decreasing lift from the wings as we reach speeds at which most tactical jet aircraft would stall. We approach abeam the ship and notify the LSO (Landing Safety Officer) that our gear is down and locked, we notify him of what the required fuel level is for our aircraft to hover, and what our actual fuel level is. The LSO, once he is satisfied our fuel levels to hover are correct and we have the sufficient engine performance to do so, will clear us to continue. At this time, our nozzles are pointing 60 degrees down, and we arm the water injection switch. In short, the water switch allows 500lb or 90 seconds of water to be sprayed into the engine to cool it off. The maximum thrust our engine produces is either limited by temperature (most likely on a hot day) or RPMs. The water switch, should the engine pass a particular threshold, will allow water to be sprayed into the engine to cool it down. As a result, we can operate the engine at faster RPMs (6–7 per cent) to take advantage of the cooler engine temperature. The end state is that should we need it, we can achieve considerable additional power to help the aircraft hover. Additional power from the water switch can be utilised both with landings and take-offs. We throttle-up the engine to full power so that the engine passes the threshold for water to flow into the engine, and check that the water gauge ticks down and the water light turns on. The LSO will note that water is flowing into the engine because the exhaust will become very noticeably black. Once the pilot has verified that water has successfully flowed to the engine, he eases back on the power so that the water ceases to flow (we need to save this water for when we really need it). We continue our descending turn so that we line up with the axis of the ship. By this time we should have conducted our landing checklist. Gear is down, flaps are in the correct position, the stop we use to help place our nozzles to a precise angle is completely aft allowing full use of the nozzles, the duct pressure shows that air is getting successfully blasted to the puffer ducts that allow us controllability in the hover, we press on the brakes and check the brake pressure is sufficient, we check that we have no warning or caution lights, we double check that the water switch is armed should we need it, and lastly we check our landing light is on. We slowly turn toward the ship with our landing spot in sight. As we get to the proper distance and the proper altitude window we advise the LSO that we are going to throw the nozzles from 60 to the full vertical position. The LSO will roger up this call and make us verify one more time that we have armed our water switch. As we throw the nozzle lever back and push all our exhaust directly down, it feels much like the aircraft is jammed on the brakes a little too hard. The pilot will lurch forward and the speed will drastically tick down. The most important and often most perilous portion of VSTOL flight is transitioning from wing-borne flight to jet-borne flight. It is a careful balancing act as we pass through speeds that, should we not be riding a column of air, would make any other aircraft drop out of the sky. Our wings gradually become useless, and all that is keeping the aircraft aloft is this column of air and the controllability from the puffer ducts.

Every Harrier pilot is schooled from the very beginning that successful VSTOL flight relies on a strict adherence to what many call the 'death equation'. The death equation is airspeed x sideslip x angle of attack. If this equation has any value other than zero, then our jet can possibly lose control. If the angle of attack is too high, our wing can stall before we can rely on the jet exhaust to keep us airborne. If we have enough airspeed to create lift on the wings, and we have too much sideslip (imagine an aircraft skidding through the air) then one wing will produce more lift than the other and the aircraft could possibly flip over. So we must have one of these values be zero to make the equation be zero. Of course, we can't always have airspeed at zero because there is the transition phase from wing-borne to jet-borne flight. Angle of attack is rarely zero either. (This is the angle at which the relative wind hits our wings.) The only value we can realistically zero out to perform successful VSTOL flight is sideslip. On the nose of our aircraft we have a weather vane that shows the relationship between our aircraft's nose direction in relation to where the relative wind is coming from (sideslip). Our goal throughout the hover is to always have the relative wind and the nose of our aircraft at parallel. As we throw the nozzles to full hover stop, and the aircraft rapidly decelerates to zero airspeed, the pilot will dance on the rudder pedals to keep the wind vane centered. Passing through 60kts, the pilot checks the engine performance to make sure that as we lose the final bit of lift from our wings, we have enough engine performance for the aircraft to rely solely on exhaust gas to keep us airborne. If not, we must make a hasty transition back to conventional flight before we drop out of the sky. The aircraft slowly approaches zero airspeed alongside the ship's landing spot. The LSO clears us to transition sideways above the ship. Aboard the ship, there is a system of lights that cue the pilot to his exact fore, aft, and vertical position above the intended point of landing. The pilot can easily figure out his side-to-side position above the spot using the 'tram line' that extends along the longitudinal axis of the ship. In simplistic terms, once the pilot is cleared to land, he uses the lights to guide him down to a precise touchdown point. As he nears the deck the LSO will call the pilot to pull the throttle back to idle. The pilot will pull the throttle back, put his feet on the brakes, place the nozzles back to aft, and secure the water switch.

The taxi back is very much similar to the taxi out. We conduct engine shutdown procedures, record pertinent engine and aircraft data for the maintenance crew, and post-flight the aircraft. During post flight, we are generally looking for damage and missing panels. We check for things such as leaks or any damages/nicks to the engine blades. After a quick post-flight, we will go straight to intelligence, while the mission is fresh in our minds, and debrief what we saw, what we attacked, and the damage we inflicted. The Harrier can record everything seen in the head up display and what was seen in the targeting pod. We will use these videos both during the intelligence debrief and the flight debrief. For the debrief, we always begin with safety of flight. In other words, did anything happen during the flight that was deemed unsafe? Next, we will debrief the administrative portion of the flight. We generally concentrate on what we did wrong or what we could have done better. After this, we concentrate on the tactical portion of the flight. This is where the HUD and targeting pod video come in handy. Almost every second

of the video is critiqued with the same questions in mind. What did we do right? More importantly what did we do incorrectly? How could we have done things better? The flight lead will write up a matrix with his individual mission goals and discuss (with check marks or X's) whether they were successfully accomplished. Much like the brief, the debrief is covered extensively (especially the tactical portion which is debriefed by the second). Following the debrief, the wingman will type up a report of lessons learned during the flight and send it to the flight lead who will then add his thoughts.

This sums up a typical flight in the Harrier Community. As stated before, the Harrier is also capable of flying an assortment of missions outside of CAS. The overall planning, execution, and debrief occur in quite the same fashion though. I would lastly like to add that the Harrier mission is constantly evolving. Though it is deemed as an older generation aircraft, the platform has done an outstanding job in maintaining state of the art technology – specifically our targeting pod and advanced weaponry. As we maintain our technological prowess, commanders in the field still discover many new uses for the Harrier's unique capabilities.

5

PIRATICAL PERFORMANCE

During the 1980s, apart from the defensive tasks undertaken by both the Phantom and Tornado F3s (in company with Hawk support), the RAF also maintained a large number of offensive aircraft, such as the Tornado GR1, the Jaguar and the Harrier. Following the outbreak of hostilities in the Middle East during January 1991, the RAF began to operate some of these attack aircraft 'for real' over Iraq and occupied Kuwait, and for the first time since 1956, the RAF was again engaged in the business of warfare.

The press initially speculated that the Gulf operation, codenamed Desert Storm, would be a swift and decisive conflict, but the Pentagon and the British Ministry of Defence held a more realistic view from the outset. Crucial to their plans were the continual bombing sorties flown by RAF Tornado and Jaguars against the heavily defended Iraqi positions. These attacks were pressed home despite the danger, and in the knowledge that crews were being shot down.

To the outsider, however, one initially surprising development was the decision by the British to send Buccaneers to the Gulf, little more than a year before their scheduled retirement. The reason why the 30-year-old Buccaneers were pressed into active service during their twilight years was the specialist weaponry that they carried, and the equally specialised missions that were flown by their crews. Laser-guided bombs (which can be aimed precisely on targets with a laser beam) could already be carried by both the Tornado and the Jaguar, but the Buccaneer crews specialised in 'laser designation' – the illumination of targets with their ingenious 'Pave Spike' pods.

The RAF afforded me the opportunity to fly in a Buccaneer and I eagerly accepted, knowing only too well how rare such opportunities were. The Buccaneer is a two-main aircraft and the second seat was not designed for a mere passenger, therefore the occupant has to perform at least some of the duties that would normally be undertaken by an RAF Navigator. Consequently, very few civilians were allowed to sample the Buccaneer's capabilities first hand.

Situated on the shores of the Moray Firth, Lossiemouth is a bleak landscape of grass and heather, broken only by ample stretches of concrete and a scattering of both huge hangars (once used by the Fleet Air Arm) and smaller HASs (Hardened Aircraft Shelters). It is one of the RAF's busiest airfields, currently providing a home for Tornado bomber squadrons, but also acts as a temporary base for an endless line of visitors destined for the nearby weapons ranges and low-flying areas, where the RAF and other NATO units are permitted to fly below their usual 250ft height limit.

During the 1980s, Lossiemouth's two HAS complexes housed the Buccaneers of Nos 12 and 208 Squadrons, which formed the basis of the RAF's anti-shipping strike force. The two units were assigned to SACLANT (Supreme Allied Commander Atlantic), operating exclusively in the maritime environment as part of No 18 Group. Like any other front-line

ABOVE: A line-up for the prototype NA.39 Buccaneers pictured at Holme-on-Spalding Moor. As can be seen, the aircraft wore slightly varying paint schemes and national insignia, all designed to frustrate attempts to establish the aircraft's size through espionage. (*British Aerospace*)

squadron, 208 had its own dedicated complex of briefing facilities, shelters and support facilities, all 'hardened' against NBC (Nuclear, Biological and Chemical) attack, and surrounded by a wall of barbed wire, guarded by the RAF Regiment during exercises (when wartime conditions are simulated). The various buildings were divided into the 'hard' and 'soft' areas (i.e. those that are strengthened to withstand attack, and those that are not), and the Buccaneer crews moved freely to and from each area. Buried in this very private citadel, Buccaneer operations were seldom seen in any detail by any outsider, and although visitors to the two resident Buccaneer squadrons were common, only a few very lucky individuals were given an opportunity to see Buccaneer operations from the closest perspective – the cockpit.

The first task any potential Buccaneer 'back-seater' must perform, civilian or otherwise, is to become familiar with his working environment. Lossiemouth had its own Buccaneer simulator system, an authentic representation of a real two-man cockpit, providing an ideal place for a newcomer to get to know his temporary 'office' a little more intimately. There was no such thing as a free ride in a Buccaneer, as there were some important jobs (albeit very simple ones) that had to be done by the occupant of the navigator's cockpit (i.e. tasks that could not be performed by the pilot). The main responsibility for any passenger was the control of the fuel supply. On the forward position of the left instrument console is a small control panel with five toggle switches and three corresponding dials. These switches control the fuel flow from the two underwing tanks (if installed) and the bomb bay tank, as well as further reserves housed inside the rotating bomb door itself. It was this last facility that gave some variants of the Buccaneer its distinctive bulged belly appearance. The fuel has to be fed into the aircraft system as required, and the gauges have to be constantly monitored. If the appropriate switch isn't thrown at the right time, the fuel lubrication system could overheat and catch fire – not the kind of situation that any fast-jet pilot would like to find himself in. Directly opposite the fuel supply controls, on the right forward console, are the controls for the IFF (Identification Friend or Foe) and SSR (Secondary Surveillance Radar) system. Most RAF and civilian airfields utilise SSR, allowing radar controllers to keep a constant tally of aircraft call signs, heights and flight paths, and it was the navigator's responsibility to control this function if

required. On the main forward instrument panel are the drift variation switches for the pilot's bomb-aiming facility, with a set of weapons station selection switches on the side of the left-hand cockpit wall.

The crowded cockpit's most noticeable piece of furniture is the radar CRT scope, wedged into the left-hand side of the instrument panel. Usually fitted with a tubular shade, the mode, brightness and contrast controls for the display are placed immediately to its right. The main radar controls (including the Off, Standby and Active selections) are over on the right console. There are one or two other items of interest too, such as the VHF/UHF radio, the altimeter, compass, speed indicator, and the radar altimeter, just visible on the pilot's instrument panel through the clutter that occupies the void between the front and rear cockpits. It takes a couple of hours to become fully confident in knowing just what piece of equipment is placed where in any combat aircraft, and

ABOVE: A look into the Buccaneer's cockpit, illustrating the twin-seat arrangement and the blast screen designed to protect the navigator/observer. The ejection seats were fixed either side of the aircraft's centreline, giving the occupant of the rear seat a slightly better forward view. (*British Aerospace*)

the Buccaneer was no exception, but after some familiarisation the initially forbidding 'black hole' looks a little more friendly.

The need to dress appropriately for Buccaneer flying means a visit to the flying clothing section, where suitable kit is issued and coaxed on, sometimes with considerable effort. First come the 'long johns', huge woollen socks and a roll-neck sweater, over which the familiar flying overalls are worn. The weighty boots rarely, if ever, fit without a struggle. Over one's legs and stomach go the anti-g trousers (referred to as the 'gsuit') and with the aid of clips and long zip fasteners, this inflatable structure is pushed and pulled into place, before being tightened to individual size by means of various adjustment laces. The aircrew helmet is the next item to fit, and with a struggle the assisting NCO adjusts a suitably sized item to sit 'comfortably' on one's head. The fit is tight, and it has to be so, as the fast-jet environment (high-g conditions) would soon shift a loose-fitting helmet, and the impact of an ejection would render a badly fitted helmet dangerous rather than beneficial. The oxygen mask is then attached, and this too has to make a tight fit, not only to ensure a good supply of air and oxygen but also to safeguard against any poisonous fumes that might invade a cockpit. The mask also forms part of an overall facial guard against bird strikes, which can smash a canopy and blind aircrew if not protected. Many aircraft are also fitted with MDC (Miniature Detonating Cord) systems, the Buccaneer being no exception, and therefore the possibility of an accidental canopy shatter is also always present, leading to fragments of Plexiglas and molten lead flying around the cockpit. After checking that the RT lead and microphone (fitted into the mask) are functioning, and that the head set fitted into the helmet is also performing loud and clear, the clothing can be stored away until it is time to fly.

As well as making these mental and physical preparations to fly in a Buccaneer, passengers in such aircraft are required by the RAF to be certified as 'fit' by the RAF's medical officers. You don't have to be Superman to qualify, but equally the RAF does like to be fairly confident that a passenger is going to survive the experience in one piece, and great care is also taken to establish whether the occupant will be sufficiently small to fit into the confines of the rear cockpit.

The next day starts early, and an MT section minibus calls at the station entrance to carry personnel to the 208 Squadron shelter complex, a five-minute drive away, around the airfield perimeter. The first call is to the crew room, where the duty pilots and navigators are sampling the endless supply of coffee, tea and biscuits. The wall-mounted television is tuned into the BBC's Breakfast Television, and the assembled air and ground crews divide their attention between their refreshments, the TV news, and various magazines, newspapers and station information sheets. The scene is a comfortable one, and there are few uniforms here, other than the traditional drab flying overalls. Occasionally a pilot walks by carrying a 'bone dome', life jacket, and anti-g trousers. His journey will lead him just a few yards from the crew room to a locked door, which will require the insertion of a code number to allow access. Immediately next to this door is another much larger version – an enormous steel monster of a door with a huge locking wheel. Moving this door requires considerable effort, so it is generally left open during normal peacetime conditions.

The floor inside this entrance is not solid. One walks along the corridor on a raised metal grille, a few inches above a pit containing fuller's earth. In here, the surroundings take on a more serious appearance – we're in an airlock. Following the one-way system, thoughtfully signposted by arrows (wartime NBC contamination regulations dictate that the wearers of 'dirty' clothing follow just one route into the building, and 'clean' personnel another route out), one enters a second squadron crew room, this time deserted, with a dormitory visible to one side, and a kitchen and washroom opposite (used only during special exercises that replicate full wartime conditions). No time for rest or refreshment here, as the next room is the 'business end' of 208 Squadron's operations – the all-important Ops room, containing all current information on the unit's day-to-day operations. The wall carries a large visual display which shows the status of each of the unit's Buccaneers (weapons fit, location, etc.), the schedule of missions for the day, crews involved, times, fuel states, and much more. To the left side is a large diagram of the HAS complex. In the centre of the room an SNCO engineering controller sits at a long console, in company with an Ops control SNCO and a senior squadron executive officer. They are effectively running the day's operations, in direct connection with all sections of the squadron and the outside world. During exercises (and in a real conflict) the 'War Executive' would sit here and co-ordinate 208's operations.

Through an internal window, the Buccaneer crews can be seen, hard at work inside the planning room, a larger area containing two spacious tables, map stowage shelves, and wall-mounted information connected with day-to-day flight planning. The room is invariably littered with maps of all descriptions, as the pilots and navigators deliberate over high- and low-level routes, drawing up mission maps with the aid of pens, rulers, protractors, compasses and pocket calculators. The adjacent door leads to the briefing room, where the pre-flight briefing for the next sortie is ready to begin. It's going to be a two-aircraft exercise, intended to include almost every aspect of routine Buccaneer operations. The four men involved in the operation (including the author and the Commanding Officer of the squadron, referred to as the Boss) prepare to digest the appropriate details from the mission's appointed leader, who stands at the front of the group beside an overhead projector:

Okay, good morning, gentlemen. The aim of today's sortie is to go off as a pair and use some weaponry, and do some low-level navigation, with a main aim of doing Buccaneer-to-Buccaneer air-to-air refuelling. We also have a VC10 tanker as a bonus, but everything is aimed at getting the Buccaneer tanker. So write down the details please, and look up when you're ready.

The crew members glance up at the projector's wall display and scribble the important points on to their knee pads, fitted into their flying overalls. The morning's weather details have already been covered, leaving the officer to explain the mission plan:

Okay, I'll carry on now. Crews are myself and Mike leading, with the Boss and Tim flying in number two. Call sign will be Skull One and Two, and the HASs situation is with one-six-eight in twenty-four, and you've got two-eight-seven in HAS twenty-two. We have the SX fit and you have the CE. Your load is two inerts, probably on station three, and two flash stores on station four … drop the inerts, please. We have the standard stuff with three KGs. We have Eighteen-K of fuel for ourselves, and nineteen for you, Boss.

From this barrage of abbreviations the crew learn that the sortie is to include the use of some underwing stores (in this case two inert bombs), some low-level navigation over land, and some AAR (Air-to-Air Refuelling), using both a VC10 and another Buccaneer, equipped with a 'buddy' refuelling pod. The call sign to be used for radio communication for the formation is 'Skull', and the aircraft (XV168 and XT287) are housed in HASs (Nos 24 and 22). The fuel loads are 18,000 and 19,000lb respectively, and the underwing stores are fixed in SX and CE configurations (code letters for two specific combinations). The brief continues:

Weather here at Lossiemouth is quite nice, looking out of the door. Steve will go out to help you and set up all the switches, Tim, so there should be very little for you to do, other than fuel management and looking after yourself, okay? Then you can start up and do the checks as you require, Boss, just to be there in time for the check-in. We'll taxi out to whatever runway is in use, and it's one-zero at the moment, and we'll line up as per the wind. We'll use a twenty-second stream and I'll be going blown, then you chase after me Boss to catch up. Join at whatever formation you require, and if you need any photos, Tim, just tell the Boss what you want. I'll try and go for Arrow formation, but if you want to come in close, please make sure I know so that I don't rack on the bank too much. However, whatever you like, Arrow or Battle. En route to Rosehearty we will use the formations I've already covered, whatever you require, Tim, for your photos, just put in your request to the Boss, and he'll clear it through me, so I don't go and do anything silly and bash into him! When we've completed the sortie, we'll come back in here for a pairs approach. On one-zero it's an SRA, and hopefully on two-eight a PAR or we may just do it visually, whatever we feel like at the time. Just make sure you're upwind of us, Boss, and we'll do hand signals. We'll overshoot the pairs approach at four hundred feet. I'll put the airbrake in, and there'll be a pause. When the airbrake's fully travelled I'll gently start to bring the power up, so the airbrake's your signal for when I'm about to start overshooting. We'll try to keep it nice and tight. Whoever is towards the downwind leg will turn downwind away from the other guy, into a positive turn, and get themselves in the circuit. Landing separation is twenty seconds minimum, and remember there will be other people's circuits as well, so you won't get cleared to land inside that anyway. All through the exercise, if we go unserviceable you'll have to wait until we get ourselves another aeroplane, and if you go unserviceable, we'll wait for you.

The odd phrase 'going blown' refers to the Buccaneer's BLC (Boundary Layer Control) system, which blows high pressure air (bled from the engines) over the control surfaces, thus creating greater lift and allowing the aircraft to fly at slower speeds than were necessary for carrier approaches (the Buccaneer being built to naval specifications for the FAA). Normal Lossiemouth departures were performed 'un-blown', but for this sortie the main (8,393ft) runway was temporarily closed, and the shorter (6,023ft) secondary runway was to be used instead. SRA (Surveillance Radar Approach) and PAR (Precision Approach Radar) are different types of ground controller-guided approach patterns used as required for day-to-day flying. The navigator now continued the briefing:

Time check … we've had Royal Flights and there's one to the south so it shouldn't affect the sortie, but we've had to sign for it. Danger areas … there are two that affect the sortie, eight-oh-nine south which is active to eight thousand feet, so we'll transit above that on our way out to the north, and six-oh-nine, plus these over here which are suppressed throughout Exercise Priory, so we need to keep out of them. Were we for some reason needing to work to the south here, we would certainly need to keep out of six-oh-nine, six-oh-seven and six-oh-eight for Priory. Safety altitude, we can use five-point-five throughout for this part of the high ground, and remember there are lumps of rock sticking out of the ground around there. Notams … there are none that affect us, although there is a late warning that I'll point out when we outbrief, of a crop sprayer. Airways, and again we'll be transmitting over the advisory routes, but we'll be getting a radar service from Scottish. If our internal chatter is too much I'll ask for a frequency that they don't mind us blathering about on. Coming back low level we should be under all those. Pressures we'll take when we walk. Fuel is one at ten, and

probably the same after the tanking, or Bingo, minus whatever it is after the tanking, so that we know after all the different tanking events going on, where we start from. Bingo two of five-point-five is a fuel from about the Orkneys, to come back low level through the Pentland Firth, a fairly straight recovery on five-point-five. Okay, the plan then … if it's two-eight, a big right turn, or if it's one-zero, straight ahead, more or less heading zero-nine-zero out towards Rosehearty. It's five minutes from abeam the lighthouse, so we'll be in at the target at four-twenty knots. We'll go in Battle with you on the left, and when we're abeam we'll turn downwind and increase to five hundred knots and squawk four-three-two-zero. There are helicopters around there, Tim, so keep your eyes open as well, and if you see anything, talk about it and tell us. Okay, we're in there for two passes, probably visual laydown, and we'll rendezvous at Troup Head off the second pass, although you'll probably be fairly close behind us I'd guess, and you call us to turn, jink or whatever, right? We'll then set off round this little low-level navigation stretch … We'll try and fly this route, and you feel free to move around us as you see fit for a good line for your photos, Tim. The only thing to beware of on this route is Balmoral out here to the right. That'll take us to Montrose and we go under this airway here as well, so if there's weather down here, the base area level is six-five. And that'll take us to Montrose … measure the heading … looks like zero-eight-zero from Montrose, climbing up to two-three-zero initially, and we'll go across to Scottish then. We'll call when clear. To RV with the tanker, call sign India India eight-five. We'll climb out heading up into the towline talking to Scottish, and then hand over to Buchan. Then we'll endeavour to do the VC10 tanking. You can take pictures, we can manoeuvre, or do what we like … they're very flexible, the VC10s, and they'll probably do what we ask for. We'll wait with them, hopefully until the Buccaneer tanker turns up, call sign India Echo Uniform three-three, from his Tornado task, okay? He'll try and get there between fifteen hundred and fifteen-thirty, so he should join us on the towline, and he'll probably take a bit of fuel as well, so we can both watch that, and then we'll set off, heading up towards Duncansby Head, to do our Buccaneer-Buccaneer tanking. So we'll leave Buchan, go back to Scottish, try and get a quiet frequency, and do our Buccaneer-to-Buccaneer tanking on that leg. The tanker will go home and we will let down to low level, heading about north from around the Pentland Firth, to look at a tactic … and we'll do half a Delta. Simon will brief it. Then we'll follow round the Orkneys and if we've got the fuel and time, we can wander around there as much as you like. There are light aircraft around the Orkneys, so keep a look out for those. And then back through the Pentland Firth, past the oil rigs back to Lossiemouth, okay? We'll commence recovery on the Bingo two call … Any questions?

The fuel calls are again mentioned, chiefly because things are far from straightforward when aerial refuelling is included. The pre-briefed fuel figures obviously have to be updated, depending on the amount of fuel taken by each receiver. The phrase 'squawk' refers to the IFF/SSR system (Identification Friend or Foe), describing the act of transmitting an information pulse to the ground receiver. 'Buchan' and 'Scottish' are area radar controlling authorities. 'Priory' is the name of a three-day RAF exercise taking place at various RAF fighter stations, involving the use of a variety of reserved airspace. The pilot now takes over the briefing again:

First task then, we go to Rosehearty and drop our bombs. Now, you've already been shown the bomb distributor, I believe, Tim, okay? There aren't many switches you need be too worried about on there, apart from the Start, Stop and the Singles switch, and we'll cover those fully at the end of the briefing. Rosehearty range then … we'll do two passes with the twenty-eight inerts, and the three KGs for us. Rosehearty range is the call sign and stud seventeen. Target two, which is a raft and as you'll see it's a small orange raft that sits out there in the middle of the water. Sometimes you can see it and sometimes you can't. Sea level and danger area SOP, and what we're really looking for there is to put the bomb close to the target, so you don't bomb any fishing boats, and the bomb ought to drop between a thousand foot over and a thousand foot short. The Boss will make sure he's quite happy that the bomb's not going to hit anything else, and then he'll call singles; that is your cue, Tim, to make the final safety switch selection from Off, and it should already be Start-three

and Stop-three, okay? That's providing the inerts are on three, which they ought to be, but you'll check that in the book when you get out there, Boss, two-five-one is the heading that we aim to drop the bombs on, and what they give you the score off. We'll go for a visual split, and make sure you're a good twenty seconds behind, and we'll RV at 5 nautical miles north of Troup Head … and as soon as you're aboard call us, and we'll head off on our route. The wind we've got … we're looking at a slight southerly drift of fifteen knots, but really at the bombing height it's plus ten, and your own speed is five hundred and four knots, and we'll work our own out. Now Tim, you want five-point-one set to allow for the weight and the type of attack you're doing, and the correction of the aircraft. Forget about wind velocity, the Boss will offset for that … if you lay forty feet into the wind for every ten knots of wind, a quick look at the wind, and just aim offside. The raft is thirty foot long as a gauge itself. We're looking for the bombs inside one-forty feet. Once you're complete on the range you must make the switches safe, Tim, and we'll make a switches safe call to remind everyone to do it, so remember you're setting Start-eight and Stop-one, and the mode switch goes to Off, and the Boss will make his safe.

The Singles, Stop, and Start references are connected with the weapons station selection and release switches, which are controlled from the rear cockpit, although the final signal to fire was made (as in most two-crew aircraft) by the pilot. The briefing continues:

Okay, the low nav bit. We'll run along there and all sit on track, but if you want to do anything, Tim, like taking photos, just say, otherwise sit back with the Boss and he'll give us some cross cover. If we do hit bad weather at this stage, remember that there are solid centres in those clouds, so get away in an emergency and abort from low level. You know how to squawk emergency, Tim? If we end up going into the airway we will tell you, but obviously we'll try not to do it. Happy with that? All right, we'll coast up, and then go up to the tanker. First of all then, it's India India eight-five, a VC10 at fifteen hundred to fifteen thirty on the two line. Buchan will get a frequency for them, and expect us to take Tacan five-eight to get a range from him. He's planning to be flight level two-five-zero, and we'll join with an RV Charlie using full RT. If you spot the tanker before us Boss, then try and get our eyes on to it, and we'll get ourselves aboard nice and swiftly. If we need the airbrake I'll take it, I won't call it. We'll join on the port in Arrow initially, and we'll refuel in the order one and two, and we'll take two-K apiece. Have you been to a VC10 before, Boss? Well, if we've got a half-hour slot it might be worth taking a few dry prods as well, but if our tanker has called up we'll just take two-K and get him in and fuelled up. We'll depart with our tanker to Duncansby. We'll call the VC10 and when we speak to him with our requirements, we'll tell him that we have a press photographer in the back of number two. While we're in that state we will hopefully have been joined by India Echo Uniform three-three, and he will come in and tank. We will have gone through first, and we'll initially sit off to starboard and wait for him to finish. When he's complete and dropping out, we'll start to move out into tight Battle formation so that he can move up on the inside. We'll have passed on our requirements on leaving, so we'll turn off there. And then it's back to Scoots … hopefully not in that fashion there … [the magnetic Buccaneer shape being used to illustrate the positions suddenly performs its own dramatic dive from the blackboard, accompanied by hoots of laughter]. So, on to Duncansby, flight level as required, probably around the two-fifty mark. Once clear we'll try to do it as silently as we can, so as not to clutter up the Scottish frequency, but if you want any inputs, Tim, speak up and tell the Boss what you want and he'll tell us. We'll tank in the order one and two, initially for one-K, but if you want us to take dry prods and stuff, we'll change headings to get the sun out of the way, just tell us, as at that stage we'll have plenty of gas. Once we're complete we'll be on the starboard and we'll clear to low level. Emergencies at any stage during the tanking are SOP the diversions, and you can go to Leuchars, otherwise back to Lossie. If you have a problem we will come with you and drop you off somewhere, but if we're with the Buccaneer tanker, we'll let him take you home, and if we've got a problem we'll go into Leuchars. The VC10, in case you can't remember, Boss, has quite a lot of dihedral which is quite disorientating, if there's a bit of cloud around, but otherwise it's quite

smooth, and it's a nice big basket which tends to wobble a bit … they're very good. Missed approach then is as standard, obviously you've probably got something wrong, so return to your stabilised position, check trim and have another go. If there's any damage to your aircraft, then it's into Leuchars, and we'll come with you. Right, are there any points on refuelling? It's the main reason for going up there anyway. Now the tactic … to make it nice and simple we'll do a Delta Two experimental, one and two in the fifteen hundred yards swept position. We'll use two hundred feet to get it away from the sea, and please watch the weather there. Then climb out … if the weather's funny, then try and get yourself into close formation and ease away from the water, and if not we'll just pull up, call our heading, and you take us through. Escape manoeuvre as SOP, and use instruments in the toss recovery. Once you're happily back on the horizon get me visual and follow me out. Switches, well if you like you could do a vari-toss but you won't have the offsets … follow me up using 4g. To let you have an idea what we're talking about, Tim, we're simulating attacking a ship that is probably undefended, and we're dashing in to toss a bomb at it from about 3½ nautical miles, okay? So what we do is at thirty miles we go up to five-fifty knots to get there nice and quickly as we come over the horizon, and at fifteen miles we'll call 'Bananas' – a ridiculous thing to say, but what it means is that the target is at 15 miles on the nose. We'll dash in, and at 3½ miles we'll pull up. You will hear us say 'Standby … now' and then you'll hear 'beep' which is us going up, and five seconds later the Boss will have started his stopwatch and will follow us. After four seconds the bombs would come off, heading towards the target, and the Spikers, the guys down at the back, would begin Spiking as you saw in the films earlier. We'll recover back the other way using fourg, and providing you're visual, Boss, just call us and we'll begin doing our next business.

The term 'Spiker' was central to the world of Buccaneer operations, where the Paveway bomb and Pave Tack designator system became a valuable part of the RAF's inventory. Inside the briefing room the Buccaneer crews were delighted to show visitors their video films of the aircraft in action, using the Paveway LGB (Laser-Guided Bomb) attack system. A 'Spiker' was a suitably configured Buccaneer, carrying a Pave Spike laser designation pod on the port inner hard point. Once a target was identified, the pod was used to illuminate the aiming point for the LGBs, carried by accompanying Buccaneers or Tornado GR1s. The bombs were thrown into a huge parabolic trajectory in a 'toss' manoeuvre, and once the bombs had reached the upper limit of their flight path, the laser pod was brought into action, 'lasing' or 'spiking' the target. The bombs then locked on to the illuminated target, and fell with astonishing accuracy. One film showed a tiny wooden splash target being obliterated by a concrete-filled inert LGB: 'We had this little raft with a couple of bed sheets tied to it … we split it into two parts with an inert LGB … it's a very reliable system.' The briefing is nearing its conclusion:

Right, fuel management: Tim, that's your main job on this sortie. The switches have all been explained, but when we're on the tanker you must switch the overloads off. Your bomb bay will probably be empty so switch the bomb door off … and also you can use the radar at any stage during the sortie, but put it on standby while we're tanking, so that you don't irradiate the poor guys. And pins … Steve will be with you when you strap in, to make sure they're all out, but when you come back in, before you unstrap, make sure you've got them all back in. There's no rush to get out unless it's on fire, in which case, run like hell! Generally then, Tim, on cockpit management, you've got plenty of kit with you so be careful where you put it, but at all times, if you're moving things around, make sure you can get to the ejection handle easily and quickly. If the Boss takes a bird in the face at low level, and he can't speak to you, you've got to be able to get out of there fairly quickly, otherwise he's already briefed you on what's going to happen if you need to get out. Disorientation is also an issue, so if you feel funny and you don't know which way up you are, just tell the Boss, and he'll tell you what's going on … hopefully! All right, we've covered everything else, and for a low-level abort in emergency, just get away from the ground and we can sort things out at leisure. Right, now everything on this sortie should string together, however it doesn't take long for things to start going wrong, or for people to start cancelling, so we'll stay flexible and we'll talk about what we need on the radio once we get airborne. Are there any points to add Mike? Boss? Tim?

Briefing completed, the crews leave the darkened room to gather at a small window looking into the Ops room, where any last-minute detail changes or confirmations are raised (known as 'out-briefing'). It is now time to gather the remaining items of flying clothing equipment. The lifesaving jacket is a particularly cumbersome piece of kit, but it serves the wearer well, and is standard to almost every RAF flying unit, often being worn even on sorties that do not include flights over water (there are plenty of lakes around where a life jacket would be just as vital). Inside the jacket is a radio homing beacon that automatically transmits a signal that can immediately be picked up by the RAF's rescue helicopter units, and also included is a set of miniature flares, designed

ABOVE: An eye-catching line-up of Buccaneers with wings folded at Yeovilton. These aircraft are Buccaneer Mk 1 variants, identified by the smaller engine intake required for the Gyron Junior engine. The overall white paint scheme is indicative of the aircraft's primary nuclear strike role. (*Tim McLelland collection*)

ABOVE: Buccaneer S Mk 1 XN928, illustrating the all-white paint scheme that was initially applied to the Navy's Buccaneers, indicative of the nuclear strike role for which they had been designed. (*Tim McLelland collection*)

machine is a former FAA aircraft with many hours of carrier-borne operations under its proverbial belt, but after the demise of the Navy's last carrier, the aircraft was transferred to the RAF and soon resumed its maritime attack duties from Honington, before moving north to Lossiemouth. Two Buccaneers are regularly housed in a single HAS, although in wartime conditions only one aircraft would be placed in each HAS, protected against the ravages of whatever offensive action might be unfolding outside. Inside the gloomy shelter, the Buccaneer appears to be huge. It's not a small aircraft, and the bulge of its ventral fuel tank combined with the oddly shaped 'area-ruled' fuselage, is enough to give the aircraft an even greater impression of size and weight. The Buccaneer looks every inch a bomber. Climbing up the access ladder into the cockpit when laden with flying clothing, helmet and a camera is itself something of a feat. Some careful footwork allows the back-seater to step over the side of the fuselage and carefully stand on the ejection seat pan, before slotting one's feet down on to the floor and then sitting back, whilst holding on to the instrument panel coaming. Once settled as snugly as possible (ejection seats are not known for their softness and comfort), the first task is to connect the PEC to a socket situated on the right-hand side of the seat. This rubber-tubed attachment supplies oxygen to the mask and also pumps high pressure air to the gsuit. On the other side of the seat is the PSP clip, which attaches the wearer to the dinghy pack (slotted into the base of the seat). Also fixed to the seat is a small window containing a weight figure dial, and this is adjusted to match the seat occupant's all-up weight, in order to give the seat's rockets the correct ejection trajectory to ensure a safe exit.

to attract the attention of one's rescuers once they (hopefully) arrive in the vicinity of an accident. Attached just below one's knees are leg restraint garters, which are suitably adjusted so that the pair of buckles stand proud, to accept a pair of lanyards that are connected to the ejection seat. If the seat is fired the lanyards retract into the seat base, snatching in both of the occupant's legs to prevent them from flailing about, smashing against the cockpit walls, and being broken. Finally, a pair of white kid leather gloves is donned to complete the pre-flight preparations. Not only are the gloves comfortably thin (to allow the effective operation of the aircraft switches) but they are fireproof too. It's time to go flying.

Lossiemouth's HASs are arranged in a distinctly untidy fashion; there are no neat rows here but instead a deliberate arrangement that is cunningly designed to thwart a marauding bomber's attempt to destroy more than one HAS on a single bombing run. Inside HAS No 22, its doors already open, is Buccaneer S Mk 2B serial number XT287, waiting with an accompanying ground crew. This particular

Next come the leg restraints, these two blue cords being threaded in turn through the twin leg buckles, before their plugged ends are inserted into a pair of sockets at the base of the seat. They have to be attached now, as it's almost impossible to reach them when fully strapped in. The seat and parachute straps are pulled up from between one's legs and from over one's shoulders, all routed and attached to a central quick-release box. Once each strap is successfully connected, the complete harness is tightened (each strap in turn) until it is almost impossible to move one's body. For most flight conditions, even during quite severe manoeuvring, the straps don't need to be painfully tight, but the wearer can never be certain that his flight will not end with an ejection. If an emergency egress is necessary, even an inch or so of strap slack can (and sometimes does) result in broken bones, as the seat's brutally swift progress from the cockpit makes no concession to the fragile occupant. Finally, the big, bulky helmet is pushed and pulled into place, and the oxygen mask is clipped into position. The clear visor is slid down, and now the back-seater is ready to fly, already hot and slightly sweaty inside the confines of a cramped cockpit, hidden inside a dark shelter. The combined smell of oil, fuel and rubber mask does nothing to improve the slightly nauseous feeling that tends to develop in such alien surroundings.

The Boss arrives in the HAS and, after consulting the ground crew, takes a walk around the airframe, searching for any obvious signs of potential danger. The ground engineers have already thoroughly inspected every inch of the aircraft, but nobody gives up the chance of a last-minute glance, just in case something has been missed. He then climbs into the cockpit and begins his pre-start checks, his actions visible from the back seat, thanks to the slightly off-centre seating arrangement. Rather oddly, the rear seat is positioned just to the right of the centreline, with the pilot's over to the left, allowing an over-the-shoulder view of activity in the front cockpit. Outside on the shelter floor, the ground crew wind the Palouste air starter into action and the HAS suddenly fills with noise which is muffled only by one's flying helmet (the ground crew wear ear defenders and work either by RT connection or hand signals). The Buccaneer's two mighty Spey turbofans then slowly rumble into life and the noise becomes even more overpowering as the rush of air from the starter unit is replaced by the more substantial sound of the Speys, with waves of hot kerosene vapour rolling around inside the shelter. The ground crew attending to the start sequence monitor the operation of the flying controls, signalling (by hand) to the pilot that each movable surface is operating correctly in response to control column inputs. The ejection seat pins are then removed from each seat and stowed in their flight positions. The sound of our 'playmate' aircraft and crew (call sign 'Skull One') crackles into the RT; they too are ready to taxi and with a signal from the ground marshaller, a brief blast of power

from the engines enables the huge bomber to lurch forward, making a brief curtsey as the brakes are stabbed to check their function. As the Buccaneer rolls out into the daylight the canopy is slid shut and locked, and the environment suddenly becomes rather more friendly, the roar of engine noise subsiding and the foul smell of oily exhaust being replaced by a slightly (but only slightly) less unpleasant stink of rubber, courtesy of one's oxygen mask.

Skull One slowly rumbles past ahead of our aircraft's nose, heading for the runway, and as Skull Two, we carefully tuck in behind them, on to the long 50ft-wide taxiway, with HASs to either side of us. In the relative calm, the RT chatter switches between the two aircraft and the control tower, to a persistent background of regular hisses and crackles. The crew's microphones tend to amplify the sound of breathing, which itself requires some adjustment in order to come to terms with the uncomfortable oxygen mask and the 'on-demand' oxygen system. The two 'doll's-eyes' indicators attached to the left console wink from black to white, in time with one's inhalation, and it's reassuring to see that both occupants of this aircraft are at least breathing. The control tower gives the two pilots air traffic clearance to take off, and with a last-minute update on the wind speed and direction, the two Buccaneers roll out on to the white-painted 'keyboard' markings at the threshold of Lossiemouth's runway 10, where a row of sparkling lights runs down each side of the concrete and off towards the horizon, although in the distance (some 6,000ft away) the end of the runway is visible, and it doesn't seem to be as far away as one might imagine or hope when a Buccaneer is about to get airborne from it. But the Boss is a veteran Buccaneer pilot and he knows that even with a full fuel load the Buccaneer is perfectly capable of achieving a routine take-off from such a modest runway.

A last check is made to establish that everything is in order, and the first Buccaneer, over on the left-hand side of the runway, begins to edge forward, accompanied by a cloud of dirty brown and black smoke. The engine blast from the aircraft gently buffets against our fuselage as our own engines wind up to full power. The sound is familiar to anyone who knows the Buccaneer – the harmonised whine of the two Speys gradually winding up to a shrill song that then changes into a deep and powerful thunder. The Buccaneer is ready to go, held on the brakes, everything shaking and shuddering. Twenty seconds elapse on the stopwatch (following from the first aircraft) and the brakes are released. Ahead of us, Skull One is still visible, just above the horizon, trailing a brown exhaust plume. We are now beginning to speed along the runway with steady acceleration that is barely perceptible. No neck-breaking afterburners to expedite our progress, but as the Speys continue to spew their thrust rearwards the 1,000ft markers either side of the runway roll by with increasing rapidity. The end of the runway is very visible now and swiftly getting even closer as the Buccaneer's wheels gently ease from the concrete and tuck smartly into their raised positions, accompanied by a few random clunks and a couple of gentle thuds as the landing gear doors slam shut. By this time we are already crossing the runway threshold and although the aircraft is barely airborne, we're out of the airfield's boundary and still low down in the weeds until speed increases a little more and we begin to gently climb and turn left towards the coastline, skirting the small town of Lossiemouth, which sweeps by beneath our

port wingtip. The airspeed is already up to the briefed 420 knots as the trees and scattered buildings give way to sea, with the weapons range a few miles distant. The Buccaneer has an enviable reputation for smooth low-level flight, and the first few minutes in the back seat confirm that the ride really is indeed exceptionally comfortable. The first-time occupant of the back seat of the Buccaneer inevitably expects the usual fast-jet rough ride, but the aircraft doesn't exhibit the usual tendency to bounce and dance in the turbulent air at low level. It sits firmly on course and battles ahead, seemingly oblivious of the conditions that we're flying through. Skull One (XV168) has already assumed a loose battle formation, and it can just be seen down on the horizon, a couple of miles to our left out in the gathering gloom and mist. Rosehearty weapons range controllers have already cleared both aircraft into its area, and our speed begins to rise as the two aircraft gently descend to just 200ft. With the airspeed needle now indicating 500 knots, the Boss turns the aircraft through 180 degrees towards a tiny raft target out on the coastline as XV168 dashes in ahead of us.

RIGHT: A Buccaneer S Mk 2 about to take off from Yeovilton. Visible are the external fuel tanks, carefully designed to fit the contours of the Buccaneer's wing, thereby reducing aerodynamic drag and improving performance. (*Tim McLelland collection*)

LEFT: Buccaneer S Mk 2 high over Aden, carrying SNEB rocket pods under its wings. The Buccaneer's bolt-on refuelling probe was carried by almost all aircraft, although RAF aircraft assigned to Germany did not require it, their anticipated sorties being relatively short in duration. (*Tim McLelland collection*)

The weapons release switches are selected, and the Boss calls to me to throw the final switch to make the small practice bomb under our wing 'live'. As the target swiftly approaches, the Boss calls out a verbal countdown, pressing the release button on his control column as the brightly painted raft rushes past beneath our fuselage in a blur. We race over the beach and in a cloud of water vapour we turn and climb. Round the range circuit for a second pass, and the two attacks are soon complete, the weapons switches placed back to their safe positions as the coastline rapidly approaches again. We pull up to a modest 1,000ft to overfly the small coastal towns, and then the two aircraft slowly edge together to form a tighter formation for an overland navigational exercise, descending once more into the gloomy rain-sodden hills once clear of the main areas of population.

Back to a low-level height of 250ft on a heading of 225 degrees, the two Buccaneers race towards the first turning point at Huntly, north-west of Aberdeen. The navigator's 1:500,000 map on my lap clearly shows the intended route, with one-minute intervals marked on a long black line thoughtfully penned in for me by a squadron navigator. Provided that the pilot stays on course and maintains the speed at 420 knots, this Navex should be relatively easy, the main task simply being to ensure that we maintain our low altitude but don't venture any lower. Outside of specific ultra-low flying areas, 250ft is the specified height limit that RAF aircraft cannot break during normal peacetime conditions – at least not intentionally. But 250ft is enough to ensure that crews get a good feel for the kind of altitudes that might be encountered in a wartime situation, during which the Buccaneer's ground-hugging (and radar-evading) capabilities would be put to good use. Sure enough, the next small Scottish town comes into view on the three-minute mark, as predicted by the map, and the next turn on to 196 degrees is initiated, both aircraft hauling round with some vigour, the gmeter reading just over 4g, squeezing the Buccaneer's occupants down into their ejection seats. The bleak Scottish landscape now becomes even more forbidding, and the weather is starting to close in. Balmoral should be somewhere to the right, but nobody seriously expects to see it; the royal residence is protected by a box of restricted airspace, and an intrusion into the area would be by accident rather than design. The rolling landscape gives a much greater impression of speed, and even the Buccaneer doesn't offer much comfort at low level over land, where the endless variations in terrain create wildly turbulent air that buffets and jostles the aircraft as it races along. The two aircraft have settled into a fairly routine bumpy race across the hillsides, passing all manner of sights, from tall pine trees to scattered sheep. It's something that became routine for the Buccaneer crews at Lossiemouth and a daily sight for the few people who occupy the far-flung corners of the Scottish countryside. Our route is designed to avoid overflying populated areas, but even so, the Boss keeps an eye open for any obvious places which can be avoided, and never stops keeping a lookout for bird concentrations that spell danger for an air-breathing jet at low level. The weather is still deteriorating, and a darkened grey patch looms straight ahead, which threatens to throw the formation off its intended track. As the crews know only too well, the clouds do sometimes have solid centres, and Skull One elects to fly around the weather, promptly pulling sharply away and down to the left, streaming two long condensation trails from the wingtips as the pilot hauls the huge bomber away from the approaching weather. It's stomach-churning just to watch as XV168 flips smartly over, belly-up, and drops down into a valley, under the weather, as we turn hard, back to the right, my gsuit increasing its vice-like grip as the rate of turn increases and I fight to keep my head up as the gforces pile on. Round the weather in a matter of seconds, the Boss quickly identifies Skull One again, and the formation is re-formed, the manoeuvring stops, and the world is upright again for a while.

The next turn is onto 080 degrees, and the lead aircraft is now some 200 to 300yd ahead of XT287, as both aircraft roll on to their sides and sweep over the top of a bleak hillside. The wingtip condensation trails are accompanied by intermittent sharp bursts of thick white cloud which occasionally form and puff around each aircraft as water vapour is literally sucked out of the air, over the Buccaneer's wing surfaces. The weather really is bad, but the overall visibility is still acceptable, and both Montrose and the Scottish coastline slowly come into view, the steady climb out of the low-level Navex beginning just half an hour after leaving Lossie. It's been an action-packed thirty minutes.

The ascent from the coastline takes just ten minutes and puts us into a refuelling area (a box of airspace reserved for air-to-air refuelling), roughly 30 miles east of Montrose. Our lead aircraft is in contact with the VC10 tanker which is already waiting inside the box, gently flying a wide racetrack circuit. The Boss spots the distant tanker and gently moves us into position, formating on the VC10's port wing at a height of 25,000ft. We're in the land of perpetual sunshine here, high above the clouds. The massive fin and the elegant tailplane of the huge four-engine VC10 seemingly sweep majestically over the cockpit of XV168 as the Buccaneer edges in to take a careful stab at the starboard wing refuelling basket. We are invited to take up position behind the port basket, and so we gently approach the rear of the tanker, before edging slowly forwards to the trailing basket. After a minimal amount of gentle adjustment, our refuelling probe nudges into the VC10's refuelling basket, and with a reassuring clunk our probe makes a firm connection with the fuel vent, inside the basket. Fuel is flowing and we take on 2,000lb of kerosene high over the North Sea. The Boss makes it look deceptively easy although it isn't; even though both aircraft seem to be motionless, the Buccaneer and VC10 are speeding through the sky together in perilously close proximity at more than 400mph, and even the slightest miscalculation could result in disaster.

Buchan radar directs a third Buccaneer into the refuelling box, this aircraft being configured as a tanker, carrying its own hose drum unit (HDU) in a pod, attached under the starboard wing. XV868 had already been hard at work further south, refuelling a formation of Tornados, and now it is the turn of the Buccaneer tanker pilot to take on some fuel for himself from the much larger reserves held on board the VC10. The Boss asks the VC10 captain (who is responsible for controlling the four-ship formation) if we can leave the standard refuelling formation, to adopt some rather more artistic angles for the author's camera, and the VC10 crew are more than happy to oblige, so while the Buccaneer tanker replenishes its fuel supply, XV168 sits on the end of the second basket, performing a few 'dry' prods (linking up, but not taking any fuel) as practice. The Boss meanwhile takes XT287 into a whole variety of different positions, looking at the formation from each side, as well as from below and above, while my camera clicks away. Keeping a receiver plugged into a basket requires a great deal of concentration, especially over a long period, and for Skull One this rendezvous must be something of a marathon; it is doubtless with some relief that the Buccaneer tanker completes its refuelling exercise. The VC10 captain departs for the long flight back to Brize Norton, turning south and gradually disappearing atop the fluffy cumulus clouds, leaving the three Buccaneers to turn on to 314 degrees, descending back to low level.

Under control of Scottish radar, the next leg of the sortie takes the formation north-west towards the Orkneys, during which time a further refuelling rendezvous is made, this time between the two receiver aircraft and the Buccaneer tanker. XV868 trails a long hose and basket from its wing-mounted Mk 20 refuelling pod, and Skull One and Skull Two each take 1,000lb of fuel from the tanker while the formation makes steady progress towards the Orkneys at a speed of 280 knots (the speed being dictated by the fuel hose's limiting speed of 290 knots). Once complete, the tanker navigator reels in the basket, only to find that the last few feet of hose refuse to slide back into the pod. This isn't an unusual occurrence, and some careful airspeed control enables the Buccaneer's wing flaps to be momentarily extended; this gives the hose a useful aerodynamic nudge, and the basket quickly pops back into the end of the pod housing. Descending slowly, the three Buccaneers level out at 250ft, just off the north-east tip of the Scottish coast, sweeping past the famous John o' Groats landmark. The tanker pilot calls that he still has a little fuel in reserve and offers to fly in formation with the basket extended once more, to allow the author to gather a few more interesting pictures. The lowest permitted refuelling height is 2,000ft, so no attempt can be made to take on fuel at this low level, but the Boss decides that we don't require any more fuel and closes in on the port side of the tanker for some more photographs as XV868 dashes along the coastline of the Orkneys, the extended refuelling basket waving and waggling in protest at the non-standard (and very turbulent) altitude at which it is deployed.

After passing the famous 'Old Man of Hoy' stack of rock, the tanker departs the formation and sets course for Lossiemouth, while XV168 pulls into line-abreast for a second pass along the coastline. The assembled fulmars and gannets that frequent the coastline have just settled on their nests again after the first disturbance, but now the two smoke-billowing monsters come hurtling round the clifftops and scatter the sea birds again, leaving their wake of thunderous noise that rolls across this normally tranquil stretch of scenery. Immediately to port is a wide expanse of water called Scapa Flow, a reminder of military conflicts of the past and the fact that this area hasn't always been quite so peaceful.

As if more proof were needed of the Buccaneer's outstanding capabilities, the final phase of the sortie begins over the Pentland Firth, north of Lossiemouth. The plan is to fly a 'Delta Two Experimental' – a technical phrase that describes a two-aircraft simulated Paveway attack that will represent the type of operations performed by Buccaneers during the Gulf War. Although today's demonstration won't include the release of any live weapons, the manoeuvres flown will be exactly the same. If this was for real, additional Buccaneers or Tornado GR1s

would be trailing us at a distance of about 4,000–5,000yd, carrying Pave Spike designation pods, but today we are unaccompanied. The radar altitude reads just 200ft, and the speed has risen to 550 knots. We're really shifting along and the cold, grey sea surface seems to be so close that it feels as if it is about to wrap itself around the Buccaneer's canopy. On an operational attack, at about 40 miles from the target, the formation splits into two elements, with either two aircraft selected as leads, or four if the formation is a six-ship. The two Spikers carrying the laser designated pods always fly at the rear of the group. The aircraft are all widely separated, allowing each Buccaneer crew to maintain an effective lookout on the neighbouring aircraft's six o'clock position, from where an attacking fighter is likely to make his strike. In this sort of formation it's quite possible to find an enemy fighter closing in on the tail of the lead aircraft, only to become sandwiched in front of the Spiker, which will normally carry a Sidewinder missile. The Sidewinder is a very effective piece of self-defence equipment, although another very useful Buccaneer tactic is to 'drop one's knickers' in self-defence. This bizarre phrase describes the release of a 1,000lb parachute-retarded bomb in front of an attacking fighter, a measure almost as effective as a Sidewinder from the rear: 'Dropping a thousand-pounder in his face should certainly put him off his aim.' About 25 miles from the target, one of the aircraft will quickly pull up to locate and nominate a suitable target feature, giving the bearing and distance to the rest of the formation, before dropping back to wave-top height. In peacetime the attacks are usually flown down to 200ft, although sometimes as low as 100ft, and even this meagre altitude can occasionally be transgressed. The wartime height limit is in fact only constrained by the pilot's confidence and nerve, and 50ft or less is a more realistic figure for the final phase of an attack, as a pilot describes: 'You might well fly the last few miles at 50ft, but you can't keep that up all day, it's too exhausting, keeping the aircraft right on the sea surface.' If a Nimrod reconnaissance aircraft is in the vicinity, the formation relies on a VASTAC (Vector ASsisted TACtic) to locate the target. The VASTAC requires the Nimrod crew (flying at higher altitude) to pass the location details to the Buccaneers, thus avoiding any risk of the Buccaneers (which would otherwise have to use their own radars, and give away their location) being detected by enemy radar.

We are now thundering towards our target, speeding over the turgid, grey sea at breathtaking speed, but even as we thunder ahead, the mighty Buccaneer remains steady, with just an occasional bubble of turbulence throwing us momentarily up or down, left or right. On the call of 'Bananas', our lead aircraft turns hard to the left and we turn with equal aggressiveness to the right. Skull One pulls steeply into a climb, and we follow just four seconds later, the Boss hauling the big bomber into a fast climb that results in more hard gforce and yet more clouds of vapour blossoming around the cockpit exterior. The steep fourg climb squeezes the pilot and navigator (or in this case an observer) down into their seats, and the climb then develops into a steady roll, after which the bomb release point would be reached and (if weapons were being carried) the Buccaneer would be releasing its deadly load of laser-guided bombs. Pulling smartly down again towards the horizon, XT287 heads back for the relative safety of ultra-low level, while Spikers would now be illuminating a suitable spot on the side of an enemy warship, to which the bombs would relentlessly be heading. The two LGB carriers drop back to the wave tops to make good their escape and the attack demonstration is over. Although the Paveway is essentially a good-weather system (the bombs need to see the illuminated target) it's a very effective means of destruction. It's important, however, that the target isn't Spiked until the LGBs have reached the top of their launch trajectories, otherwise they will prematurely begin to home in towards the target without sufficient energy to get there (their toss climb takes roughly seventeen seconds). At about 8 miles the two Spikers turn to present their port wings to the target and the navigators bring the laser beam on to target by means of a pistol grip and thumb wheel, and a TV screen. The camera head of the Pave Spike pod also provides a very useful visual recording of the attack, which can be replayed at debrief.

The Paveway system does require the aircraft to make fairly close approaches to the target, at ranges as short maybe as 3 miles. It was therefore very much a secondary means of attack for 208 Squadron, and their main offensive equipment was the BAeD P3T Sea Eagle sea-skimming missile, a 'fire-and-forget' weapon with all-weather day and night capability. There were many differing attack profiles practised by the Buccaneer crews, all culminating in what was probably their most awesome deterrent, a six-aircraft 'Echo-One' Sea Eagle attack. In this profile, six Buccaneers would each carry four Sea Eagles, flying line-abreast, less than 100ft above the sea surface. A Nimrod providing a VASTAC or SURPIC (SURface PICture) would give basic area information on possible targets, or if necessary, one Buccaneer would briefly pull up and designate a target. Once fed with the appropriate information, the awesome twenty-four-strong battery of missiles would be released. The Sea Eagle was ECM (electronic countermeasures) resistant and could be launched at distances in excess of 68 miles from the target. After firing their missiles, the Buccaneers would turn through 180 degrees and depart, unseen by the enemy. Travelling at Mach 0.85, the missiles would quickly close on the unsuspecting target ship: 'With a target like a Kirov class cruiser, there is a need to saturate the defences, but they can't pick off twenty-four missiles at once.' Total destruction was guaranteed, as the target vessel would have been virtually defenceless against these unseen assailants, armed with overpowering force.

Heading back towards Lossiemouth, the Boss elects to demonstrate a typical operational wave-top dash, dropping our Buccaneer to just 100ft, at which height it looks and feels as if we are riding the wave tops in a boat, rather than skimming above them. There is a sensation of speed but without any land features to look at, it's impossible to accept that we are in fact travelling at more than 500mph. It's as good a demonstration of a Sea Eagle attack profile as one can get. There were no specific attack manoeuvres for launching this missile, as it was capable of turning into the target from virtually any position. During peacetime practice Sea Eagle attacks, the Buccaneer crews would fly ASMD (Anti-Ship Missile Defence) profiles: 'We follow the kit to see where it leads us … we pretend actually to be the missiles.' Thus the unsuspecting target ship (usually belonging to the Royal Navy) got an opportunity to train its gunners and missile operators with attacking aircraft. If the Buccaneers had been doing this for real though, the targeted victims would have never even seen their attackers. Obviously the Buccaneer presented a much bigger target than the actual missile, and flew at a higher altitude (10ft being realistic for the Sea Eagle), but the approach speed could be imitated with precision, which gave the naval forces a useful training opportunity.

As the tower at Lossiemouth gives us the latest weather details, XV168 pulls into tight formation on our port wingtip, extending the large petal airbrakes under the aircraft's tail, which take six seconds to open fully. Their effect is most impressive, especially at high speed. ('It's a bit like flying into a brick wall.') From a transit speed of 420 knots, we're suddenly down to an approach speed of just 180 knots as the landing gear slowly extends, rumbling and clunking as it unfolds. The Buccaneer's landing is almost always flown 'blown' (i.e. with Boundary Layer Control air pressure applied to reduce landing speed) and consequently the engine power is maintained at a fairly high setting in order to keep the high-pressure BLC in operation (speed being checked with the airbrakes). The control surfaces are brought down to their 45:25:25 positions (reflecting the corresponding instrument readings in the cockpit) and the Buccaneer assumes what is often referred to as 'marginally stable flight'. With typical aircrew understatement, this means that the Buccaneer is not, perhaps, the easiest of aircraft to handle in the airfield circuit, with lift directly dependent on the engine power. The Boss comments: 'She's not a happy aircraft with flaps and undercarriage down, on only one engine, so it's important to keep a careful eye on what is happening when you're in this configuration, just in case a control problem emerges, or if you happen to lose one of the engines.' Fortunately the Buccaneer rarely needs to operate on one engine, as the Speys are remarkably reliable: 'The Speys are excellent. They rarely exhibit any problems and they're very tough. They can eat birds and don't even notice them.') As Skull One turns on to finals ahead of Skull Two and drops down on to the concrete, the second aircraft lines up on to approach. The ADD (Airstream Direction Detector) provides a visible and audible indication to the pilot of the aircraft's AOA (angle of attack) and the whine and whistle of the Spey engines begin to rise as the noise of airflow slowly gives way to engine thrust and the almost musical sound of the two Speys being gently coaxed into providing the necessary bursts of power that are necessary, weighed against the use of the airbrakes. The navigator also hears the ADD through his helmet, and after a final burst of thrust from the Speys to keep us airborne, the bulky Buccaneer thumps onto the concrete as the ADD tone changes to a noise of protest – definitely not a welcome noise unless the aircraft is already squarely over the runway.

ABOVE: NA.39 XK523 illustrates the carefully designed proportions of the Buccaneer as it sinks below deck on the hangar lift. The hinged wings and nose cone, together with the petal airbrakes, enabled this relatively large aircraft to be accommodated on the Navy's small (Second World War-standard) lifts. (*BAE Systems*)

Some three hours after departure, XT287 is back on the concrete at Lossiemouth. Already the ground crew will have been notified of our arrival and will be awaiting the aircraft's return at the HAS. XV168 is already back in the squadron complex as we turn off the runway and on to the taxiway, our canopy sliding open to allow the welcome fresh air to flood the cockpit. The safety pins are all replaced in the ejection seat, apart from the face-blind handle pin, which is out of reach (the ground crew will take care of this). The radar is switched off, having been used only briefly on this long flight, as the back-seater was preoccupied on this occasion with other matters; in any case the subtleties of the display would doubtless have been lost on an untrained civilian's eye.

LEFT: The Buccaneer's upper surfaces, illustrating the wing fold mechanism which is exposed to view following the removal of the spring-loaded doors that were originally fitted to cover the area. These doors proved to be troublesome and were eventually removed from the fleet. (*British Aerospace*)

Back inside the HAS complex, the engines stop, the pilot and accompanying observer release their harnesses, PECs and PSPs, untangle the leg restraints, unclip the RT leads, and confirm that the last pins are in place. The Boss stays in his 'office' while the ground crew connect a winch to the Buccaneer's fuselage and pull the aircraft backwards into the HAS, ready for refuelling and preparation for another sortie. Finally, crews return to the PBF to complete a speedy and informal debriefing, and the sortie is completed.

The Buccaneer was a remarkable aircraft that performed very effectively, and with its deadly Sea Eagles it presented a formidable prospect to any would-be aggressor. This capability, together with its Paveway equipment, made its deployment by the RAF in the 1991 Gulf conflict a great deal less surprising than some commentators implied. The crews, of course, matched the impressive capabilities of their aircraft, flying almost every day at high speed and low level over land and cold, lonely seas. They made the job look easy, but it was incredibly hard work. Pilots have literally flown into the water before, and others may well do so in the future. But the two-man Buccaneer crews took a philosophical attitude to the realities of their job, though the pilots and navigators certainly didn't fool themselves that they were in the joyriding business. As a mere passenger, preparing to walk out to the Buccaneer and see for myself what it was all about, I was asked by a pilot if I was nervous. I said that I wasn't, having become quite accustomed to fast-jet flying. His response was unexpected and surprising: 'Well, I always am when I'm going to do something dangerous.' Naturally, I ventured to ask if he was joking, or if he really did think that Buccaneer operations were not the kind of activity that should be regarded with flippancy: 'Well, I don't know … I suppose it is. Yes, it is dangerous actually … bloody dangerous!'

6

HEAVY STUFF

If there is such a thing as a typical RAF pilot, he is probably the chap you see on the recruiting posters, posing beside his Typhoon fighter. Of course, it can be easy to forget that the RAF does not just fly fast jets. One only has to think of the relief flights to Africa, or the long Nimrod sea searches over our stormy seas, to realise that there has always been more than one type of RAF pilot. Leaving aside the world of helicopters, the RAF's inventory of aircraft can be divided into two main categories: fast jets and multi-engine types. During the 1980s, the former group consisted of aircraft such as the Tornado, Harrier, Phantom or Buccaneer, whereas the latter group encompassed the bigger 'heavies' such as the VC10, Nimrod, Hercules or TriStar. Almost every would-be pilot has a desire to fly one type of aircraft or another, and, as one might expect, many have a desire to become a Tornado pilot, or something similar.

Unfortunately, despite the RAF's continuing requirement for such people, circumstances often prevent many flying students from becoming fast-jet pilots. There are those who do not work effectively on their own, and there are those who equally cannot work as part of a two-man team (essential for the Tornado candidate). Some cannot cope with the high-speed/low-level environment, and some simply would not physically fit into the confines of a jet's cockpit. Thus, many students find themselves unable to continue training as a fast-jet pilot, and a proportion of these candidates were routinely taken from the fast-jet training 'stream' and placed in the multi-engine equivalent.

Consequently, one might be forgiven for thinking that the prospect of multi-engine flying is second-best, but of course this certainly is not so, particularly in more recent years when the RAF's role has changed to embrace many more types of support missions. Some candidates choose 'heavies' as a first option, as there are always students who have a desire to see the world and enjoy a more varied flying career rather than simply pursuing the fighter or bomber role. Obviously the pilot of a VC10 or Hercules inevitably saw far more exotic places than a Jaguar pilot could have ever imagined. After completing their first 100 hours of basic flying training on the Jet Provost, the RAF's students were placed in one of the two categories. Those chosen for the fast jets were posted to 4 Flying Training School at Valley, where they would eventually continue their training on the high-performance Hawk. The multi-engine students took a different route and were posted to RAF Finningley (now Doncaster Sheffield Airport) in South Yorkshire, where the Multi-Engine Training Squadron (better known as METS) was based.

Finningley was one of the RAF's biggest and busiest stations, with a whole variety of aircraft types making use of the station's excellent facilities. The navigational trainers (Jet Provosts and Dominies) used by 6FTS were the most common types, but the Sea Kings and Wessex helicopters of Nos 22 and 202 Squadrons were also much in evidence, as were UAS Bulldogs and AEF Chipmunks. The fact that Finningley boasts a huge 200ft x 9,000ft runway (built for use by Vulcan bombers) often resulted in a wide variety of visitors, ranging from TriStars and

KC135s to F15s, Phantoms and Drakens amongst many others. As I lived relatively close to Finningley, the station was almost a second home to me and I spent a great deal of time there, either attending the annual air shows, visiting the resident units, or simply lurking in the boundary hedgerows, watching (and photographing) the many aircraft that could be seen there going about their business. Naturally it was a great pleasure to have a chance to see the based aircraft up close and to talk to the personnel who operated them. It was an even bigger thrill to go flying sometimes, and I seized the chance to join the METS to fly on board one of their Jetstream twin-engine trainers.

The METS first came to RAF Finningley in 1979, when the unit moved from RAF Leeming in North Yorkshire. The unit operated eleven Jetstream T1 aircraft, which were built at Prestwick by Scottish Aviation, the aircraft having originally been designed by Handley Page shortly before that famous company ceased to be. Originally some twenty-six aircraft were ordered, as a replacement for the piston-engined Vickers Varsity, but as the first few aircraft were being delivered to No 5 FTS at Oakington and the CFS at Little Rissington, the government decided that there was no longer any need for an aircraft specifically operated for multi-engine pilot training, and thus withdrew the Jetstream from service. The few machines that had reached their bases were flown to St Athan in 1974 and placed in storage, pending a decision on their future. However, late in 1976, the Ministry of Defence had a change of heart and decided that there was a case for the Jetstream after all, and eight aircraft were removed from storage and entered service at Leeming.

Thus, the METS became part of No 6 FTS, but this association was essentially a 'paper' one, as in most respects the unit operated in isolation, even possessing a separate flight line on a disused runway (now part of the airport terminal). The students arriving at Finningley for multi-engine training came directly from basic flying training, and were placed on an AFT (Advanced Flying Training) course, which lasted roughly twenty weeks. As part of this course, some six weeks were spent in ground school, and up to forty-five hours of 'flying' would be conducted on two flight simulators. The two simulators were operated by a staff of three within the METS complex, which was situated almost in the centre of the airfield, next to the Jetstream flight line (and now long-gone, underneath what is now the airport departures lounge). One simulator was an instrument trainer, which provided an electronically operated representation of the Jetstream's instrumentation. The other simulator was a full-flight, three-axis motion imitation of a Jetstream, complete with a projected visual image on the windscreen. Whilst a very large portion of the flying training was thus synthetic (almost as much as true flying) there was still no totally convincing alternative to the real thing, and so a further forty-seven hours' flying was conducted in the air, on board the Jetstream. The non-synthetic (i.e. real) flying was designed to include general aircraft handling, circuit flying, procedural flying (high-altitude and airways operation), stalling and, of course, asymmetric operation, which was one of the most important aspects of the course. Night flying was conducted on selected evenings, the four night sorties being divided into two in which the QFI (Qualified Flying Instructor) acted as captain, and two in which the student was captain, and effectively flew 'solo'. In fact, the student never actually flew on his own as multi-engine flying is always a two-man operation. These sorties included plenty of circuit flying (mostly at Finningley) together with flight direction instruction and NDB (Non Directional Beacon) homing and holding practice, which led into the procedural phase of the course. Normal daytime procedural flying was essentially the same, but use was made of civil airports such as Leeds, Newcastle, Prestwick and East Midlands. At the end of the course the student would fly to RAF Gatow in Berlin, a two-hour flight which was particularly useful in that it involved the passage through many different airways authorities, and provided plenty of practice for the student.

In order to obtain a clearer picture of how the RAF trains its multi-engine pilots, I was invited to observe a typical training sortie from Finningley as part of a current AFT course, and the I duly reported to the station on a gloomy October morning, ready to obtain a first-hand look at training in action. Out on the METS flight line, just three aircraft were visible, with two other aircraft in the air on training sorties. Normally five aircraft are brought on to the line each day, from a total of eleven aircraft that are actually available (some undergoing servicing). Each of the Jetstreams would fly an astonishing four sorties each day, two in the morning and two more in the afternoon, after a lunch break. In between the two morning and afternoon missions, a 'running change' of crews was performed, the transfer taking place whilst the Jetstream sat with the engines still purring. Our aircraft was XX496 'D', which had already returned from the first two operations of the day, and was fully fuelled and ready for the afternoon. Alongside was XX497 'E', which was also due to depart on a similar sortie at the same time. The plan was for the two machines to take off together and formate for a short photographic session, before going their separate ways.

Flt Lt Dion Hickin was the appointed pilot for the day, an instructor on the squadron, with some 1,500 hours on the Jetstream under his belt. Prior to joining METS he had enjoyed a long and happy association with the Vulcan, flying over 2,000 hours with Nos 35 and 44 Squadrons at Scampton and Waddington. The METS QFIs (of which there were six) were normally second-tour officers, with plenty of instructional experience; however, as part of an experiment, two first-tour QFIs were on the squadron. Together the QFIs had a very wide experience of almost every aircraft type one could imagine.

For instance, Sqn Ldr Fox boasted many hours on the Meteor, Vampire, Hunter, Hastings, Beverley, Varsity, F27, Anson, Valetta, Hercules and Gnat! The METS Commanding Officer, Sqn Ldr Andy Tomalin, was a former Victor pilot. Others had experience of aircraft such as the Andover, Dominie and Shackleton. Obviously this pool of practical knowledge was particularly useful when the unit had to train students for such a wide variety of aircraft types. Although the basic flying training was essentially the same for each student, once he had been earmarked for a specific aircraft, the training could be accented towards the type of flying peculiar to that type. For aircraft such as the Andover, procedural flying was clearly the most valuable training, whereas a more 'physical' handling approach was preferable for a prospective Nimrod pilot, who was likely to spend most of his career at altitudes of little more than 250ft.

Flying with Flt Lt Hickin would be Flg Off. Gavin Dobson – a typical AFT student, flying his third Jetstream sortie. On today's flight he would be consolidating the knowledge and experience he had gathered from his last two sorties, flying at various heights, performing steep turns and stalls, in both 'clean' (flapless) and 'dirty' (flaps and landing gear down) configurations. A few circuits were planned for Waddington, before returning to Finningley for a few more 'rollers'. However, before the main business of the sortie could get under way, there was to be the fifteen-minute photo-session with XX497, courtesy of its pilot, Dave Carter. After completing a pre-flight briefing, the crews prepared to depart for the flight line, but the METS Commanding Officer (or the Boss as he is more commonly known) had other ideas. He pointed out that the rule book required a fully qualified pilot to occupy the student's left-hand seat during formation flying, thanks to a rule concerning low-speed and flap-extended operations. Thus another crew member was needed, and thankfully Flg Off. Mark Pickavance happened to be passing and was cordially invited to join the afternoon's trip. He did not need much persuading, and promptly followed us to the two aircraft.

After settling into the aircraft, the two Turbomeca Astazou 16D turboprops are brought to life, each possessing some 1,000eshp (engine shaft horsepower). The noise level inside the aircraft begins to rise quite considerably, chiefly because there are no luxuries such as sound insulation in the RAF Jetstreams. The RAF aircraft (200 series

ABOVE: Jetstream T Mk 1 XX499, illustrating the aircraft's simple lines. The Jetstream was a direct follow-on multi-engine trainer for the RAF, replacing the venerable Varsity. (*M. Freer*)

machines) are essentially 'shells' with little if any concession to aesthetic appearance or comfort. However, despite their functional appearance, they are fully equipped for their training task, with three seats 'up front', the third seat being positioned behind the left-hand pilot's position with a swivel facility, allowing it to be placed centrally. Behind these is another pair of forward-facing seats, a selection of emergency equipment, and a pair of rear-facing seats positioned next to the crew entry door, on the left-hand side of the rear fuselage. Communications equipment consists of two VHF/UHF radios, twin VOR/ILS, DME, ADF and IFF, together with automatic pilot. Twin flying controls are fitted like any multi-engine type, and all the crew positions have RT connection points.

Having called Finningley control tower, XX496 is given taxi clearance and 'Foxtrot Yankee Tango Niner Three' ('Alcove Formation') moves forward from the flight line. Our aircraft is positioned 50yd behind XX497, which is slowly weaving down the taxiways towards runway 02, where Dave Carter brings his aircraft to a halt on the left-hand side of the runway, just forward of the landing threshold. Thanks to the ample width of Finningley's runway, there is plenty of room for us to park line-abreast on the right-hand side. The two-aircraft combine is then given take-off clearance and the engines are brought up to full power. Throughout the flight the Astazou runs at a constant 43,089rpm, the power variation being supplied by altering the degree of propeller blade pitch. With both aircraft literally straining on the brakes (XX497 visibly pushing down on the nose wheel) we are ready to go, '497' releasing brakes first, while Flt Lt Hickin sets the stopwatch counting on the instrument panel. The noise and vibration are, by now, stronger than ever, and it is with some relief that, with ten seconds on the clock, the brakes are released and XX496 lurches forward, surging down the runway in pursuit of '497' which is already airborne. With a forward speed of 95 knots, we have arrived at VR (rotation speed) and the nose pulls sharply upwards, the Jetstream floating effortlessly into the gloom, with airspeed at just over 100 knots, giving us the safety climb-away speed (V2). Into a left turn out towards Lincolnshire, and a few seconds later we are already well within range of '497', which is positioned ahead, to starboard. The cloud then closes in, leaving us briefly on instruments, before we ease out over the cloud

ABOVE: Jetstream XX497 lands at Cranwell after completing a training sortie. The Jetstream was withdrawn from RAF service in 2004, the multi-engine training role being taken over by Beech King Air turboprops. (*Tim McLelland collection*)

ABOVE: XX496 is now part of the RAF Museum's collection at Cosford, resplendent in its bright colour scheme and complete with No 45 Squadron's markings on the nose and tail. (*Tim McLelland collection*)

RIGHT: An impressive line-up of C-17 Globemasters pictured during an exercise at Nellis AFB, from where the aircraft flew support missions over the Nellis test range. (*US Air Force*)

tops, only to find the aircraft sandwiched between a low cloud layer and an equally thick high cloud layer, which is not going to help my photography attempts one bit! At 6,000ft we edge slowly towards '497', allowing me to train my camera on the aircraft, through the right-hand portion of the windscreen. Dave Carter then eases back slightly, and begins a steady progress to a precise line-abreast position, while I reposition myself further down the fuselage in order to obtain a better view through the starboard windows. In the interests of safety it is not possible to bring the Jetstream too far back, or too far in towards the camera ship, as Flt Lt Hickin would then lose sight of the aircraft, which would, of course, be very dangerous for all concerned. Thus XX497 remains just slightly behind, in line with the tailplane.

With photography finished it's time for a swift change to port, repeating the whole process off the port wing, lowering flaps and undercarriage, raising them again, and finally rolling away to the left in a fairly steep bank. With this complete, '497' comes alongside for one final visit, while the combine turns back from its north-easterly heading (which has taken us beyond Leconfield and out over the east coast) into a south-westerly direction. We say our goodbyes and, with a final steep bank away, XX497 is gone, off on a separate sortie, leaving us to settle down to a more typical procedure, with Flg Off. Dobson taking position in the left-hand seat, ready for the official beginning of his AFT sortie.

The first objective is to complete a series of stall manoeuvres, which requires an altitude of at least 8,000ft, so a steady climb is initiated, taking us to '12K' over North Lincolnshire at a speed of 150 knots. No specific areas are designated for stall practice, and provided there's no conflicting traffic, and a clear visibility of at least 2,000ft, anywhere will do. The Jetstream ends up quite close to Scampton, and a careful eye has to be kept on their airspace, as the unmistakable shape of five red Hawks can be seen, high above the cloud tops, performing what will doubtless be the beginnings of another superb nine-ship 'Red Arrows' formation. With the cockpit checks complete, we are about to see exactly what will happen in a slow 140-knot landing approach, with 'everything down', a situation that would occur should the turn on to finals become just a little too steep for comfort. The airspeed slowly bleeds away, passing 100 knots and downwards, until the ASI reads 90 knots, and as the airspeed continues to drop, the stall warning buzzer sounds and Flg Off. Dobson pushes the nose firmly downwards, putting the two Astazous into full power (48° pitch) which pulls the aircraft neatly out of the rather uncomfortable limbo of an impending stall. After another thorough search of the surrounding airspace, it is time for a repeat performance, and the whole sequence is followed through again, the gear coming down, flaps down, blade angle back to 15° and the speed starts to fall, 130 ... 120 ... 110 ... 100 ... 90 ... – in comes the buzzer, the stick-shaker operates, and on comes the power, gear up, and we accelerate away again, before attempting a third sequence. With these manoeuvres complete, Waddington is called, and permission is granted for XX496 to descend and join their approach pattern. Descending through the gloom once more, we eventually sink into the comparative clarity of the Lincolnshire countryside, west of Waddington. Still descending, the aircraft passes directly over their NDB, and turns left into the landing pattern, behind a Canberra. Flt Lt Hickin expresses some concern that we may well lose sight of the twin-jet in the poor visibility, but while it can still be visually tracked, there's no great problem. Fortunately it's possible to maintain contact, albeit with some difficulty, and so we continue to proceed towards the airfield. With gear and flaps down, the speed at 140 knots, Flg Off. Dobson eases the Jetstream round the left-hand turn on to finals, speed decaying to 120 knots. The aircraft is in the right place at the right height, and more or less at the right speed, but the QFI is not too happy. We are slightly wide of the centreline, and the runway threshold is fast approaching. Gradually the nose edges on to the steady line of approach lights, and XX496 slowly sinks over the A15 and on to Waddington's runway 03 at 90 knots. A slight ballooning effect follows before Flg Off. Dobson brings the Jetstream (very) firmly down on to the concrete. Rolling forward, Dion Hickin calls 'power' and the Astazous are brought up to full power, and with a very impressive surge of acceleration, the nose pulls up and we are up into the air once more passing the Falklands veteran Vulcan XM607, down off our starboard side (a position from where the aircraft was subsequently moved, many years ago). Flt Lt Hickin is immediately prompted to recall the number of times he has performed this procedure over the very same runway, but at the controls of a Vulcan. The sensation was virtually the same, but the residents of the little village of Waddington surely prefer our soft purr to the roar of four Olympus engines that once rattled around the locality. Turning left into the downwind leg, the Canberra is still in the pattern, and somewhat closer than before. Throughout the sortie, Flt Lt Hickin (the QFI) has been effectively acting as co-pilot, leaving the flying to Flg Off. Dobson, although the steady patter of instruction is maintained at all times. The student is encouraged to become accustomed to the 'feel' of a larger multi-engine type, which behaves so much differently to a single-engine aircraft. The effect of slipstream on the rudder and elevators has to become familiar, as does the associated trim changes required in response to power variations. With power on, the nose pulls up, and as power is reduced the opposite applies. Both conditions need to be 'trimmed out' almost as second nature. With power brought back, the slipstream effect is negligible, whereas at full power, the pitching effect is quite marked. Another problem to be overcome is the reluctance to 'slam' the throttles; the two Astazous have almost instant power (thanks to the constant rpm) and students (naturally) tend to 'nurse' the throttles slowly from idle to full power. This was not necessary on the Jetstream. What is important, however, is to appreciate the pitch variations caused by varying power setting during landing. Students often forget the nose-down effect of reducing the power, and last-minute trim changes are quite common. Despite this, the Jetstream was essentially a viceless aircraft and was an ideal machine for the job.

In order to gain a little more time on finals, this second roller will start from a rather longer approach, Flt Lt Hickin suggesting a distant country mansion as a suitable turning point on to long finals. Flg Off. Dobson obliges, and brings us round on to a long approach, to the east of Lincoln, the distant cathedral towers poking into the overcast sky. The Canberra is just about to touch down ahead of us, as the Jetstream levels out a couple of miles from the runway. The height is okay, but we are still not quite in the right position and the QFI decides to point out this error: 'Okay, look at our position … Are we at the right height? … Good, well are we in the correct position? … No we're not, so let's wop it over there then, and get it into the right place.' Flt Lt Hickin takes control and puts the Jetstream precisely on to the centreline, before handing back to Flg Off. Dobson. We pass over the ranks of rather wet aircraft enthusiasts standing on the A15 (we are in the middle of Exercise 'Priory') and bang down on to the concrete once more. The 'Power' call is made, and another surge of power from the two turboprops pulls us softly into the grey sky once more, turning back on to the downwind leg for another attempt at a perfect landing. 'Lights on, gear down,' Flg Off. Dobson calls out the checks as we continue downwind. 'Three greens,' he calls, and the turn on to finals is made once more, this time with the friendly Canberra still on finals. We tuck in behind as the 231 OCU jet flares for touch down just ahead of us. The approach is much better this time, but as the Jetstream passes over the runway threshold, the steady descent becomes a rather unsteady roller-coaster, as it hits the turbulence from the Canberra. The wings rock, and we pitch slightly, but Flg Off. Dobson keeps everything well under control and settles the aircraft on to the runway. 'We'd better give the Canberra a bit more space next time,' remarks Flt Lt Hickin as we complete another roller. Up once more, gear up, and time for one last stab at the circuit. Still a little unsure about the correct positioning, but the QFI says nothing, and suggests we return to Finningley for a few more rollers.

Waddington was a regular haunt of the METS Jetstreams, as was Cranwell, Topcliffe, and their former home at Leeming. Swinderby too was often used as a suitable practice diversion airfield. The choice of airfield was often dependent on how busy a given place might be, but weather was also an important consideration. For normal circuit flying a visibility of at least 5nm was required and a cloud base of 1,500ft. If the conditions did not meet the criteria, the instrument phase of the course might be temporarily substituted. Each sortie would last for approximately one to two hours, as this was widely acknowledged to be the ideal amount of time in which useful training could be performed without the student becoming tired and losing interest. There was no reason, however, why the Jetstream could not stay airborne for much longer, as it was a very fuel-efficient aircraft. Maximum fuel load was 1,400kg and at a typical cruising speed and altitude, the fuel consumption was just 200kg per hour. Gibraltar to Finningley required just 600–700kg, for instance.

Back at Finningley, XX496 joins the airfield circuit and proceeds to perform another four rollers before rounding out on to finals for the last time. Down on to runway 02, the propeller pitch lock is automatically disengaged, allowing the blade pitch to be changed to −5° and when the wheels are firmly on the concrete, a full reverse thrust of −14°, which brings us to a very short stop just halfway down the runway. In flight the pitch lock would keep the propeller blades automatically within a range of +9°/+48°. As we taxi back to the flight line, the next crew is already waiting to execute a 'running change' and with the two Astazous doing exactly that, we vacate the aircraft as the next two aircrew climb in. The Jetstream was normally occupied by just two aircrew (QFI and student), although the extra seats were taken up with passengers whenever possible, mostly in pursuit of air experience for ATC cadets. Just a couple of minutes later, XX496 rolls forward off its stand and buzzes away down to the runway for its fourth and final sortie of the day.

Although the flying element of today's demonstration sortie was over, the all-important debriefing was still to come, and the QFI immediately set forth on his task of completing a full written report on the student's performance. A copy of this would be given to the student to add to his own personal file, which contained reports on every stage of his training. These notes acted as a useful basis on which improvements could be encouraged and achieved. Each sortie would be given a rating between 1 and 6, and more specific notes would highlight particular good and bad points. The student would discuss the sorties with his instructor and make a mental note ready for the next trip. Post-flight, Flt Lt Hickin secured a seat in the METS crew room and began to write down his comments on today's flying, prior to discussing the trip privately with Flg Off. Dobson. What did he think? 'Well, his circuits were constantly too wide … he didn't pay enough attention to the crosswind effects, so his approach was off the centreline. He was on the wrong glide path at the wrong speed … he should have remembered that deploying the flaps will automatically kill off another 15 knots of the speed.' Obviously the critical QFI had plenty of comments to make, but criticism was (and is) an important part of the learning process: 'On take-off the nose was too high … in the stalls he spent too much time getting into them, mucking about looking for a good position. His taxiing is poor and he doesn't have good speed control on the ground … how many times did I tell him to brake on the runway?' On a more positive note, Flt Lt Hickin did admit that the student had been subjected to a distinctly non-standard beginning to the sortie, thanks to the author's presence, which doubtless unsettled the normal routine, and as an overall opinion he remarked, 'Well, it wasn't too bad actually … about par for the course … the kind of thing we would expect at this stage.'

Gavin Dobson would certainly have plenty to talk about at the debriefing, prior to the continuation of his flying training on his AFT course. He still had the delights of asymmetric operations to look forward to, examining and experiencing the effects of engine failure, the roll and yaw tendencies, and how to complete a successful overshoot from approach in this configuration, including the engine re-light drill. All this would be explored as part of the academic training, prior to a demonstration by the QFI, finally with the student actually 'having a go'. After this stage, night flying and procedural flying came into the course, and one low-level sortie, during which the Jetstream would be flown down to 250ft at around 180 knots, to give the student a taste of handling a big aircraft in this environment. Most low-level flights began with a procedural (airways) departure to Edinburgh, before descending into one of two low-level routes used by METS. One was a westerly route (via Prestwick and down to Leeds), with another along an eastern track. The routes were carefully chosen so as to be suitable for the Jetstream, which had a very limited visibility from the flight deck. Some fifty-six students were trained

within the AFT course each year, with about eight on the course at any given time. After completion of AFT, the students were transferred to an OCU, to convert on to their individual operational types. Three other courses were operated by METS, these being refresher training (which, as the name implies, was intended to provide training for pilots returning to flying after working on ground duties, and consisted of training rather than instruction), QFI (training six new QFIs each year), and MEXO (Multi-Engine Cross-Over), which retrained pilots who had been 'restreamed' from either fast-jet training at 4FTS Valley or helicopter training at 2FTS Shawbury. Students could be withdrawn from either of these courses for medical or simple ability reasons, and the most appropriate alternative to fast jets or helicopters was often multi-engine flying. MEXO was roughly similar to the AFT course, although the fact that the student would have already completed a substantial amount of flying training meant that the MEXO course was somewhat shorter, with only thirty hours of flying. The Jetstream operations were divided between all these tasks, with the largest proportion of flying being devoted to the AFT courses, which took up about 49 per cent of each year's flying hours; 15 per cent was taken up by the MEXO course and a further 9 per cent on QFI training. The remaining hours were used for staff continuation flying and 'other tasks'. The latter category was quite flexible, as the METS Jetstreams were used for many varied duties other than simply training students. Quite a substantial amount of flying was used to provide a limited transport capability for small loads that would otherwise require either a Chinook or Hercules. Transporting equipment in support of the Battle of Britain Memorial Flight (and of course ground crew) was one common activity, and, for example, at the time of my visit one aircraft was dispatched in support of two No 43 Squadron Phantoms which had performed at an air display at Toulouse in France. During the 1982 Falklands operations, the Jetstream fleet was kept very busy with a wide range of transport duties, transferring equipment and personnel to and from UK and RAFG airfields. All this 'non-standard' flying was used to provide useful staff continuation training.

With the debriefing completed, another AFT lesson was over. As the author left Finningley, the familiar hum of distant Astazous was still hanging in the late afternoon gloom as the two remaining Jetstream sorties came to an end and the aircraft joined the Finningley circuit. It was not one of the station's busiest days, and the Jetstreams were the sole users of the landing pattern. On some days, however, the circuit could be filled to capacity, with a mixture of Jetstreams, Dominies, Jet Provosts, the odd Chipmunk or Bulldog, and occasional visiting aircraft, not least Sea Kings and Wessex. The fact that Finningley could sometimes be so busy led to speculation that some of the resident units might eventually be relocated, but the METS remained at Finningley until the RAF reluctantly vacated the station in 1996, thanks to defence expenditure cuts. The RAF's multi-engine training resumed at Cranwell and the Jetstreams remained in use until 2003, when they were retired and replaced by a small fleet of Beech King Airs. The later type continues the training role to this day, but there's no doubt that the ubiquitous King Air fails to possess the character and history that was associated with the Jetstream that preceded it. The Jetstream was the very last aircraft to be designed by Handley Page, one of Britain's oldest and most famous aircraft manufacturers, and as such the type certainly earned a place in the RAF's history.

Today, the RAF's multi-engine training requirement is significantly lower than it was just a few years ago. With a much smaller force, the RAF requires fewer pilots, but new crews still emerge, equipped to fly the RAF's fleet of 'heavies' such as the Hercules and TriStar. The RAF also operates a small fleet of huge C-17 transports, although it is of course the United States Air Force that flies the biggest C-17 fleet by far. Lt Col Paul Shevlin (a USAF C-17 pilot) and M/Sgt Ron Dunn (a C-17 loadmaster) provided the author with a fascinating insight into the world of C-17 operations, as performed by them and their colleagues at March AFB in California:

Our bread and butter is aeromedical support to Afghanistan. Every Friday right now we have a C-17 leave here with two crews on board. It positions to Germany via the US East Coast where it picks up cargo first, and then gets into Germany on Saturday. Then on Sunday a new crew will take the cargo down to Afghanistan so that the aircraft is back in Germany by Monday morning. Then the crews swap again and the crew that stayed in Germany takes the plane again and it becomes an aeromedical evacuation aircraft so that on

Monday and Tuesday we can fly aeromedical evacuation flights back to back, possibly with another flight on the Friday, and then the two crews come home with the jet on the Sunday, using that flight as another aeromedical trip but this time directly from Germany back to the United States. Sometimes that will take us through San Antonio and pretty much always goes through Andrews AFB before eventually getting back home to March AFB. We do that every week but there's an overlap of a day or so, and usually one aircraft will arrive in Germany on the Saturday as another one comes back home on the Sunday. We have eight aircraft in all but that commitment doesn't take all of our resources, even though it's a big chunk. We also have an ICMPA weekly mission, meaning Inter Continental Movement of Patients, to take patients from overseas back from the East Coast to the West Coast and so on. So most of what we now do is aeromedical evacuation.

In terms of training, right now we do two local flights, on three days each week, these missions lasting around four hours, and this keeps us proficient in our various roles. We do some in-flight refueling, we use night vision goggles, and we will do assault landing out at different fields, doing different tactical approaches and departures, all of the normal training we require. Once a week we'll also go off station and go out to different locations around the USA, staying out for several days moving cargo around as required. I've been with this unit since 1997, so I've been here quite some time now. We got the C-17 in 2005 and before that we'd been flying C-141s and I would be remiss if I didn't mention that the 452nd also operated tanker aircraft. Until 1997, I had almost exclusively been flying C-141s through my Air Force career. In terms of overall footprint, the C-141 and C-17 are much the same, the difference is that the C-17 is very much a wide-body aircraft, and we can now take wider and higher cargo, which enables us to take more outsize equipment. Of course, the C-17 can carry much more weight, the maximum payload on the 141 was about 73,000 pounds and now with the C-17 it's 170,000 pounds. On the 141 we could carry only thirteen pallets on the cargo floor centerline, whereas on the C-17 we can carry eighteen so it's quite a difference. The C-17 is a much newer aircraft and therefore it's much better, even though I loved the 141. But that aircraft was

LEFT: High over Kyrgyzstan, a C-17 is about to take on fuel from a KC-10 extender tanker in June 2012. The C-17's refuelling receptacle is situated above the flight deck, as illustrated by the open door and yellow alignment markings used to aid the refuelling boom operator. (*US Air Force*)

Even today they still liaise with the crews and minor modifications are still made occasionally, based on asking the crews about functionality.

As explained, we embrace a wide range of missions within the roles that we currently fly. It can range from just hauling fresh fruit and vegetables, all the way through to combat missions into blacked-out remote fields. How you would brief and out-brief would directly depend on where we're operating. But most of what we do in any given week is the aerovac missions into and out of Afghanistan, with four missions into Afghanistan in any given week. If we look at one of those flights, it begins in Germany, we get an alert prior to the designated take-off time and we prepare right there at Ramstein. We grab some food to eat as it's going to be a very long day ahead of us of nearly 24 hours, so we load up with provisions and then head out to the airplane and drop everything off there. Then we head over to our Command and Control Centre and we will brief the cargo upload information as even though it's maybe a medevac mission, we will still have cargo and passengers as well as the medic crew. They will also brief us on any cargo we have that might contain any hazards such as explosives or corrosives, and we'll check that the load plan fits within the limits of our aircraft. The Air Transportation Operations Center people will already have the aircraft loaded for us so when we get out there we just have to check that it's loaded properly and secured, because once we take control of it, the pilot has to be briefed about the nature of the cargo and any hazards, whether they can be jettisoned in flight if we have to, and so on. Once we fly down range we then offload the cargo and we then become an aerovac flight so we reconfigure the aircraft, load the aircraft with medical crews and patients, and then fly back to Germany. Ramstein also briefs us on intelligence and tactics, depending on what is going on down there in theater.

very much '60s technology designed in the 1950s. You're dealing with much newer technology now with Fly By Wire, Head Up Displays, and so on. It's a very different situation. Now you're landing a half-million-pound airplane on a three thousand foot dirt runway. It's a very capable machine. You certainly couldn't do that in a C-141.

The C-17 was very much designed by McDonnell Douglas even though it's a Boeing aircraft. Boeing inherited a mature program that was derived from the YC-15 and technology demonstrator and a lot of that aircraft's capability was incorporated into the C-17. The USAF's loadmasters certainly like the C-17 and prefer it over the C-141. Getting rid of the flight engineers has given the loadmasters more responsibilities, understanding the aircraft's systems better, but that gets everyone more involved. The cargo rollers, tow systems and so on, it makes the C-17 a much better aircraft to work with. When the C-17 was being developed, I was based over at Norton AFB just over the hill from here, and the C-17 was designed down at Long Beach. Douglas used to take our loadmasters down there and literally get them directly involved in the design of the cargo compartment. They had test platforms and mock-ups, to test everything out, and they went through everything with direct input from our loadmasters and asked them precisely what they'd like to see in the aircraft.

The C-17 can make it from here to Europe easily enough but there's a fuel trade-off. At about 50–60,000lb of cargo, you go above that and you start trading off fuel for cargo. We can carry up to 245,000lb of fuel, but once you go over around 55,000lb of cargo, then every pound of cargo you put on is one pound less of fuel you can carry, so that is always a consideration. Going down range to Afghanistan the aircraft is generally flown as full as possible up to its maximum take-off weight, but this depends on the nature of the runway, the temperature, if it's wet or dry and so on. But we'll generally be up to our max gross weight, which is around 585,000lb. Taking off out of Germany, most of our weight is fuel as the aim is to ensure that we take on as little fuel as possible when we're in Afghanistan, so typically we'll land with about 100,000lb of fuel on board and we'll just top that off and put maybe another 50–60,000lb on top of that

while we are there, to get back with. Obviously we don't want to use the fuel out there unless we have to because it's far more expensive; the fuel has to be taken out there by tanker wagon and it's a resource we have to preserve. Fuel calculations are always an important part of our planning of course, and for example we're currently planning a flight to Australia, and we're looking at the weights and such. We have an 80,000lb payload so we're going all the way to Rockhampton in Australia and that's right on the limit of what we can do without aerial refueling, so we've planned a fuel stop en route, but we have lots of options that we can employ. Another example is when we flew some dolphins from Montenegro. The dolphins had a cabin altitude restriction and sometimes casevac patients also have a cabin

RIGHT: Despite being a large transport aircraft, the C-17 embraces modern technology, as illustrated in this view of the flight deck instrument panel, dominated by synthetic Multi Function Display panels. As a high-tech fly-by-wire aircraft, the C-17 is extremely safe and simple to operate. (*Jeff Eddy*)

altitude restriction so it's something we often encounter. As a result we have to fly lower and that increases our fuel consumption. With the dolphins, we made a single aerial refueling going over there and did a double aerial refueling on the way back. The terrain where the airport is situated is very mountainous so that affected the take-off weight and really we only had enough fuel to get as far as England, so we took on fuel from a tanker based there and then flew on to Canada to refuel again before taking the cargo down to NAS North Island at San Diego. But that was a good example of how we can employ aerial refueling when we don't want to stop. Obviously you want to get the dolphins from A to B as quickly as possible. Similar considerations can apply to patients that we carry. Another example is after 9/11 we were tasked with flying detainees to Guantanamo and that's obviously another situation where you don't want to stop off anywhere, so we wanted one straight trip to Cuba and that's a case where aerial refueling enables us to do that.

The Current Operations people are at the very heart of the mission planning process. It starts with them. We'll figure out the broad parameters of the mission and we'll bounce all the stuff off each other looking at what the requirements are. One thing we always look at is whether the cargo has been moved by us before and does it have all the right certifications. If it isn't, then we have to work around the requirements. When Nasa has needed to move satellites, they've brought out mock-ups to us and that's enabled us to determine whether they could be moved by C-17, what kind of equipment would be needed to handle them and so on. We moved the Mars Rover, which was built down at the Jet ProvostL and they trucked it over here from Pasadena and we flew it out to Cape Canaveral. The cargo compartment has to be kept within a certain temperature range for things like that, so if you're stopping any place, loading or unloading and so on, you have to plan to have people there to keep the environmental systems running. So it's issues like that that are addressed at Current Ops, examining the nature of the cargo, where it has to go and when, and all the other issues like those that have been mentioned. The pilots and loadmasters will look at everything and then begin to put a mission together. We load it into our Command and Control tracking systems GDSS, that's Global Decision Support System that everybody sees so they know what's going on. At some point in that process the mission will be handed over to the scheduling side of things so that they can start putting a crew together to fly the aircraft. Sometimes we don't know the crew until maybe a week or less before the actual flight but once we do, then we can look at the final details such as prior permissions that are often required to go into other airfields, and things for the crew such as accommodation, rental vehicles and so on. Realistically, the crew doesn't need to become involved until only a day before the flight; in fact it's the training missions that are more involved from a crew point of view. If it's an operational mission, then its main issues will already have been handled by the Current Ops people, but for local training missions we will plan our own requirements. For example, we might decide to head out low level, then go find a tanker, then maybe drop into the assault zone up at Travis AFB and do some assault approaches and take-offs over there, so with that the crew will get involved maybe two or three days prior to the day. We do have a requirement to fly low level in some circumstances, and we have to be capable of penetrating enemy defence systems if necessary, and for us in the C-17 that means going in low, so we do have to practice that. In terms of planning it means drawing up and printing out a route with any projected threats that we might have to encounter and avoid.

Sometimes we can get to austere locations where they just don't have someone who can do the load planning, and they'll just come out and say this is what we've got, can you take it? So we do our calculations and say yes or no and let the commander know how much cargo we will have so he can plan the fuel properly. You can see what can happen if you don't load properly, if you look at that 747 accident out at Bagram recently you can see what happens if there's a load shift. That's how critical it can be to load the aircraft properly. People sometimes ask us why we secure the cargo loads for about 3g forward shift, and then you see things like that crash and then you understand. It needs to be properly secured at all times, particularly on landing, as if we're all sitting forward and the cargo shifted forward too, it's going to be lethal. Plus in the C-17 we have assault landing capability and the brakes and reversers really work.

When we stomp on the brakes it stops, and anything not really tied down at the back is going to move. If you put an M1 tank on there it takes something like forty 25,000lb chains, it takes up all the chains we have on the aircraft to tie it down. Actually we have space for more so we could also carry Hummers or support vehicles but the one tank is 130,000lb so it's not so much its size but its weight. You can actually put two M1 tanks on a C-5 but it's a much more complicated task shoring up the load and getting on and off in the C-5, whereas in the C-17 you can literally just drive it on board. That means we can move large loads like that out into austere locations much more effectively.

The C-17 is designed as a three-man crew. You have the commander, the other pilot and the loadmaster. Generally, the commander will go to Base Ops to take care of pre-flight paperwork, the flight plan and so on, fuel loads and all that stuff. The copilot will go to the airplane with the cargo load while the loadmaster takes care of the operations to actually get the aircraft loaded. The copilot could refuel the aircraft if he is required to do it, especially in austere locations. Then he can do the pre-flight inspection, both exterior and interior, looking at all the equipment and servicing that is necessary. Then the commander will get back with the Notams, weather details and everything we need, and by then we should all be ready to come together. We bring the passengers out if we have any, and we move on from there. On the ground we have a built-in APU, an Auxiliary Power Unit, so we don't need any external help. However, the APU is a small turbine engine so it uses fuel and as a matter of policy we run it as little as we can and if we have an external power cart we will use that. The avionics are very sensitive to heat so we need the 90 kva APU or external power to keep the systems happy especially in high heat conditions. That gives us electrical and pneumatic power and the electrics will give us the hydraulics too so we can run everything off that. We can also start the airplane off the APU. We run a cockpit check first, go through everything on the flight deck, and ensure that all the switches are in the right place, and then we brief what order we will start the engines in. Usually it's one, two, three, four but sometimes we might start two at a time if we need to, especially if we're in a hurry – for various reasons – but that varies the airflows through the manifolds and it runs

the risk of a sheared starter shaft failure in the engine. The starters are pneumatic; it takes the air from the APU and bleeds it off to spin the engines. You push a big red button on the cockpit overhead panels for each engine. A light comes on to show that the starter valve is open, so you have APU air drawn through the pneumatic manifolds, down the pylon into the engine. The starter then starts spinning the engine and when you get to 18 per cent rpm you go ahead and put fuel in. Actually it's a fuel and ignition switch, and that starts the engine and there's an electronic control for the engine that takes it from there. It takes about two minutes to start each engine. The loadmaster will watch the start from the outside and watch for any smoke or fuel leaks, so we're a pair of eyes for the pilot while he's flipping the switches. For most of the time the loadmaster only oversees the load and doesn't really need to touch it, unless we're in austere locations, and because of the logistics of our operations, we often fly with two loadmasters in any case, and of course the loadmaster also acts as a flight engineer on the C-17.

The actual take-off or landing procedure is much the same for almost all conditions. Out of our base at March AFB we usually fly an instrument departure as the course and procedure is in the mission computer and it is displayed to us from there. You couple the aircraft onto the computer and it's all done automatically from there and everything is flown hands-off. The initial take-off up to 200 feet is flown manually and then you can click on the autopilot panel and it will fly it off whatever we've programmed into the computer. On take-off we have all our flight information on a Head Up Display, so you can fly a maximum performance take-off or a reduced power take-off, the latter being preferable as it saves wear and tear on the engines. A maximum take-off puts maximum heat on the back of the turbines and that causes more deterioration and engine failure so we do reduced power take-offs as much as possible and our mission computer takes care of a lot of that. Basically we need to be able to clear any obstacle on our take-off path with just three engines and that can require a lot of details to ensure that we can do that. Take-off speed varies depending on conditions. I've seen just 99 knots for rotate speed, which is pretty phenomenal for a plane of that size but it can go all the way up to 150. Typical is 125–130 knots in normal conditions.

Cruise speed is a maximum of Mach 0.825 but for fuel savings we tend to stick to Mach 0.74. True airspeed-wise that's about 430 knots. It's slightly slower than most airliners but it's a good speed and a trade-off against the supercritical wing's STOL capability that loses a bit of the cruise performance. We generally fly just above 30,000 feet for cruise. It's a Fly-By-Wire control system so the aircraft is very responsive. The plane handles very well, it's stable and it does what you want it to do. In fact it's easy to get very over-confident with it, as you saw in that C-17 crash where the aircraft was flown beyond its limits. You can get over-confident as it will generally do whatever you demand of it. Obviously it will warn you, but it will pretty much do whatever you ask it to do … until it doesn't!

7

FAST AND LOW

The Panavia Tornado has earned an important place in the RAF's history. Designed as a multinational project, it was initially regarded as a ghastly European political venture, hampered by politics and destined to become an unremarkable 'Jack of all trades' aircraft. In fact, it became one of the most potent and versatile warplanes ever to have joined the RAF's inventory. Although the Tornado was created as a medium-range strike and attack platform for the European theatre, it has soldiered on to serve in areas of conflict far beyond those for which it was designed. Now famous as the aircraft that spearheaded the first nights of the first Gulf War (flying perilously low-level attack missions against Iraqi air defences), the Tornado is now recognised as a reliable and hugely effective bomber, and one that the RAF is undoubtedly reluctant to abandon. Despite having entered service during the early 1980s, it seems likely that the Tornado will continue at the forefront of the RAF's order of battle until at least 2020 or perhaps even longer. It was in Germany that it first settled into regular squadron service. The Tornado was designed primarily as a strike/attack aircraft, capable of flying support, interdiction and strike missions across the plains of central Europe, and the RAF established a very significant force of Tornado aircraft at two key bases in Germany – Laarbruch and Bruggen – from where the following account of a representative mission was created:

The tasking for every individual Tornado sortie comes from a variety of sources, although many of the flights are self-generated within the squadron. In a typical wartime scenario within RAF Germany, the Tornado units at Bruggen and Laarbruch would receive the details of their mission from 2 ATAF's ATOC, the Air Task Operations Centre, from where an Air Task Message (ATM) would be issued. This would then be allocated to the most suitable aircrew (depending on tasks, crew availability, and other operational considerations) with input from the squadron's Intelligence Officer and Ground Liaison Officer – an Army officer who would be attached to the squadron to co-ordinate flight operations with the Army's units on the ground. Once the details had been fixed and the crews allocated, the sortie could then be planned. Firstly, the aim of the mission would be addressed, examining the nature of the target, its location, and the requirements of the ATM. In peacetime this might have been within a NATO exercise or perhaps an individual mission as part of a work-up to Combat Ready status. The next consideration was almost inevitably the weather conditions, both in the target area and at the home base, as well as en-route. Although Tornado was (and is) an all-weather aircraft, the weather conditions across Germany often affected mission planning, and for both safety reasons and the importance of mission success, the timings, routes and altitudes of a sortie would often be dictated by the prevailing conditions, which could sometimes be grim with driving rain or snow, and very poor visibility. The next stage in the planning process was to address the day-to-day technical issues that affect flying, such as the serviceability of airfield aids (navigation beacons, runway state, etc.) or wider issues such as airspace reservations, Royal flights (which precluded all flying near them) and more. Most of this information came through routine Notams with other additional advisory information such as 'Birdtam' (warnings of large concentrations of birds). With all of this information collected and digested, the sortie could then be planned in detail. Inside the squadron's operations centre (adjacent to their aircraft shelter complex) the aircrew could be seen hard at work, poring over their navigation charts, with calculators, pens, and copious notes. Much of the planning depended on the type of sortie that was required, as the nature of each target dictated the type of weapons that would be employed and the number of aircraft that would be allocated to attack it. This could mean the employment (or at least the simulated employment) of free-fall HE bombs, or smaller practice bombs designed to simulate their larger and more destructive counterparts. Laser-guided weapons also became an important part of routine missions, and for some of the RAFG Tornado squadrons, the use of nuclear weapons was also simulated, while the WE.177 'special' bomb was still part of the RAF's inventory. Each type of weapon directly affected the kind of delivery profile that would be employed, and while free-fall bombs might often be delivered via a straight run-through over a target (these bombs were often retarded with deployable parachutes), they could also be dropped from medium altitude, or thrown towards a target in a toss manoeuvre (particularly laser-guided bombs). Nuclear bombs could also be brought onto a target in different ways and although toss bombing (often described as LABS – Low Altitude Bombing System) was often employed, the WE.177 could also be delivered via a low pass over the target, the weapon's retardation parachute (and/or delay fuses) giving the crew time to escape from the target before it detonated. With all of these points in mind, the pilot would carefully prepare maps to show the projected route to and from the target, marking the route with timing intervals and with all of the relevant visual features highlighted, together with any temporary avoidance areas or other sensitive locations (the routine ones being permanently marked on the maps). While the pilot prepared the maps, the navigator would

plan the route, calculating fuel use, and booking times on weapons ranges when required, or making flight plans if the sortie would take the aircraft through an airway corridor.

The navigator would also plan other key parts of the sortie, including target splits (where multiple aircraft groups might diverge onto their own target approach paths) or other procedures such as a rendezvous with a refuelling tanker, when required. Although the latter exercise required a climb to medium altitude, most of the routine Tornado flying was conducted at far lower altitudes, usually at just 250ft, although peacetime restrictions in Germany often required a slightly higher altitude to be maintained. The navigator's planning culminated in a carefully recorded sequence, plotted into the Tornado's on-board computer. AT the CPGS (Cassette Preparation Ground Station) the navigator plotted the sortie's route on a 1:500,000 'half mil' map, placed on an electronic table. By 'clicking' his way through the route, the coordinates would be loaded onto cassette and then fed into the Tornado's computer, so that each turn and every point on the target run would be recorded as a series of latitude and longitude coordinates, ready for the crew to read from the aircraft's computer as the sortie is flown. Not surprisingly, a hard copy of the coordinates was also retained, just in case the electronic system failed, and traditional low-level (and sometimes high-level) maps would also be carried, often in the form of colour copies that could be stuffed into the crew's flying suits. The trusty photocopier was often regarded as one of the most important members of the squadron. With the planning complete, the sortie could now be briefed, and the crews would gather in an adjacent room while the sortie's lead pilot and navigator presented a detailed run-through of the mission, usually aided by a screen projector and an array of scribbled charts and notes. Although the nature of every sortie would obviously vary, a great deal of Tornado flying was quite routine, and for the squadron pilots the briefing was often a simple case of wading through only the minor details that might be unique to the sortie, such as call signs, fuel considerations, weather and so on. To speed up the process, acronyms and abbreviations became common with terms such as 'Bingo' (a fuel state radio call), 'Chicken' (fuel state at which recovery to an airfield is required), 'Divs'

(Diversion airfields, dictating the amount of fuel that needed to be carried), 'Ifrep' (In-Flight report), 'Sutto' (Start-up, taxi and take-off), 'TOT' (Time over target) and 'TRP' (Timing reference point). To the uninitiated, these bizarre terms meant nothing, but to the crews they were part of everyday parlance, designed to save time and effort. The complete sortie would be explored in detail, the crews frantically scribbling notes onto their lap pads on their flying suits as the details were explained, until everybody declares themselves happy with the plans, and the briefing is wound-up. With that task complete, it would be time to go flying.

After a last-minute look at aircraft status charts and a check that no new weather or flight planning issues have arisen, the crews would then make the short walk to their locker rooms to gather the rest of their flying clothing. The standard RAF flying overalls could be augmented by a heavy, thick and cumbersome rubber immersion suit if the sortie was expected to route over water (particularly in winter) and when exercises were under way, a variety of NBC gear would also be worn, together with the usual flying helmet, gloves, and the trusty anti-g trousers. With full NBC gear the Tornado crews tended to look more like astronauts than a regular Tornado pilot or navigator, but thankfully the more unpleasant parts of the crew's flying gear were reserved for less-routine flying, and for day-to-day missions the walk to the aircrew bus was an easy and carefree procedure. After leaving the PBF (Personnel Briefing Facility) a lowly minibus was normally available to transport the crews to their aircraft shelters, although walking was often a preferred option – an APC (Armoured Personnel Carrier) would probably have been employed in a wartime situation. In common with most RAF front-line airfields, the Tornado squadrons at Bruggen and Laarbruch each had their own shelter complex, with one aircraft housed inside each HAS (Hardened Aircraft Shelter), protected against the ravages of the German winters, and also safe from the threat of enemy attack (unless it happened to be a direct hit from a sizeable bomb, or a nuclear strike, of course). By the time that the crews reached the HAS, the shelter doors would inevitably be open, the ground crew having already been at work for some time preparing the aircraft for flight. The mighty Tornado, laden with bombs and fuel tanks,

inevitably presented an impressive sight, bathed in orange-yellow floodlighting, wings spread and a blaze of shelter lights twinkling above the open HAS doors. Almost immediately, the navigator made his way to the Tornado's rear cockpit, climbing the ample access platform ladder and easing himself into the aircraft's commendably spacious rear cockpit. Once inside, he could begin the careful process of starting up the aircraft's systems, particularly the Inertial Navigation System and computer, feeding in the details of the planned sortie through the prepared cassette. While the navigator got busy in his cockpit, the pilot would talk to the ground crew, checking whether any minor serviceability issues affect the aircraft before accepting the aircraft as fit to fly and signing the aircraft log book documents. With this task complete, the pilot then would make a final walk-round inspection of the aircraft, checking that the external ordnance is properly secured with safety tags removed, and looking for any potential hazards such as foreign objects or damage. The ground crew always ensured that the aircraft was already fully checked but a last look from an independent pair of eyes was always a wise move. The pilot then joined the navigator on board, climbing from the access platform and stepping onto the front cockpit ejection seat pan, before lowering himself onto the seat and beginning the strapping-in process. The power-off checks are the first on-board tasks (ensuring that no systems will operate unexpectedly when power is applied) and then electrical power can be fed into the aircraft from a ground power generator inside the shelter, plugged into the Tornado's fuselage. Once the aircraft's systems are all on line and functioning correctly, the crew then wait for other crews in the sortie to radio that they too are ready to go and, at the pre-designated time, the Tornado's two RB.199 engines could finally be powered up.

The engine start suddenly turns a relatively quiet shelter interior into a fume-filled crescendo of noise and heat, but inside the darkened Tornado cockpit the crew are protected from the developing roar by their tight-fitting helmets (and ear cuffs) and the closed cockpit canopy. Although the engines can be heard, they create only a low-level rumble, over which the hiss and crackle of the radio keeps the crew in contact with both the other crews and the airfield ATC.

ABOVE: A Tornado GR Mk 4 streaks overhead its home base at RAF Lossiemouth in Morayshire. Lossiemouth is the RAF's main Tornado base, with further squadrons located at RAF Marham in Norfolk. (*Royal Air Force*)

The duty Ground Controller gives clearance to taxi and with a call between each of the Tornado pilots, the aircraft slowly edge out from their shelters into the damp gloom of a typical German morning. Suddenly, the grey and featureless HAS complex comes alive as the gaggle of Tornados edge from their shelters and slowly weave their way out onto the main airfield taxiway, emblazoned by flashing navigation lights and undercarriage-mounted taxi/landing lamps. The short journey to the runway provides a good opportunity to run through the final pre-flight check lists and by the time that the aircraft reaches the runway holding point, the aircraft is ready to go, the crew having also removed the last safety pins from their ejection seats so that they can, if necessary, abandon the aircraft instantaneously if anything should go wrong. The Ground Controller hands over to the Tower radio frequency and the crew are given the latest wind direction and speed status, followed by clearance to line up and take off. For most routine departures two aircraft could be positioned line-abreast for a formation take-off, unless cross-winds dictated that a staggered departure might be safer. For larger packages of four aircraft or more, a series of pairs departures would be employed, or a series of individual take-off runs, separated by ten seconds or thereabouts. Lined up on the runway, the lead pilot signals when he is ready to wind engines up to full power, this final act often being timed to a matter of seconds, in line with the pre-briefed sortie details. Up comes the engine power and the gentle rumble of engine noise swiftly develops into an urgent roar that rattles through the airframe, pushing the aircraft nose-down onto its nose wheel. A slight change in the tone of the all-embracing noise signifies selection of reheat, and as the two turbofan engines spew long 20ft trails of diamond-studded flame, the wheel brakes are released, and the nose raises slightly as the Tornado rolls forward and begins to accelerate along the runway. From inside the aircraft things seem relatively calm despite the fierce roar of the engines, but from outside the airfield is now reverberating to the sound of an ear-splitting, tearing blast of engine power which, it must be said, seems far more impressive from outside the aircraft than inside it. Acceleration is brisk, even with a heavy load of external stores, and with the navigator calling 'Sixty knots … eighty knots … one

hundred … one thirty … one fifty … one fifty-five … vee, rotate' the nose raises and almost imperceptibly the Tornado lifts smoothly off the runway and begins to climb, the landing gear quickly getting tucked away with a gentle bump from the undercarriage doors as they snap shut. Out off the port wingtip, the second Tornado is also airborne, bobbing and waving gently as the pilot completes his take-off whilst maintaining a comfortable formation with the leader. Reheat is cancelled, but there is no feeling of deceleration as the aircraft is now already speeding beyond 250mph and a steady climb out from the airfield is under way, out into the dark carpet of cloud that hangs over Bruggen.

The nature of each departure varies greatly according to the destination; the weather, the number of aircraft present and other factors influencing each individual mission. Many sorties from Bruggen are flown to the United Kingdom, to take advantage of the lower (250ft) minimum height limit available in non-restricted British airspace. These flights are normally hi-lo-hi profiles, involving a high-level transit to and from a low-level route, requiring a

ABOVE: Low over the Scottish moorlands, a Tornado GR4 from No 13 Squadron thunders towards its target during a training sortie. The aircraft's wings are partially swept back in a typical cruising position, providing a good compromise between speed and fuel economy. (*Tim McLelland*)

SID (standard instrument departure) from Bruggen, flown under instructions from various air traffic authorities along the route as detailed in the flight plan filed during the mission planning stage.

Once cleared through German and Dutch airspace, the aircraft enter the United Kingdom ADIZ (Air Defence Identification Zone) and descend to low level. The aim of most peacetime missions is to locate and destroy a simulated target, and the hours of planning will be wasted if the target attack is not carried out. If the weather permits, a low-level visual formation will be flown en route to the target, maximising mutual cross-over cover to thwart an enemy attack from the six o'clock position (many sorties include a simulated attacker, in the form of a 'bounce' aircraft, which launches separately from Bruggen). Thanks to its accurate terrain-following radar (TFR) system, the Tornado can fly automatically at 250ft. Inside the GR1's radome the TFR dish projects a monopulse beam ahead of the aircraft, nodding above and below the Tornado's flight path. The TFR projects an imaginary curved line (often referred to as a 'ski toe' shape) ahead of the aircraft and if any obstruction intrudes into the ski toe the TFR commands the Tornado autopilot to climb until the obstruction is avoided. The ride comfort can be adjusted to fly a fairly smooth course over the terrain, but for a more war-like ride (better suited to enemy evasion) the TFR will haul the aircraft up and down each hillside fairly aggressively if necessary.

On the approach to the target area, the navigator will update his navigation equipment with radar fixes and regularly monitor the fuel state, engine instruments, oxygen and electrical and hydraulic systems. Likewise great attention is paid to the weapons systems as the target gets closer. At the pre-briefed split point, each aircraft will separate on to different headings, sweep the wings back to 67 degrees (they are more likely to have been set at 45 degrees during transit) and accelerate towards the target; each will arrive over the target from different directions, separated from the next aircraft by a few seconds. One of several types of delivery profile may be flown, each being relevant to the type of weapon being carried. Laydown attacks require a direct overflight of the target and its defences – a risky tactic, but a necessary one for the delivery of some munitions, including the JP.233 airfield denial weapon used so

successfully during the Gulf War. Conventional 1,000lb bombs can also be delivered in this manner, but the resulting shallow angles of impact will generally reduce quite significantly the amount of damage caused. Loft attacks, involving a pull-up toss release several miles from the target, allow a generous stand-off distance from the target, avoid a potentially dangerous overflight and also give a larger angle of impact for the bombs, which thus may cause much greater damage. The RAF Tornado force would also use this profile to deliver nuclear free-fall bombs if required, although the WE.177 nuclear bomb (assigned to RAFG Tornado squadrons) could also be delivered in a laydown mode, thanks to its built-in retardation parachute.

A shallow dive attack will give the aircraft some protection from being detected. The Tornado approaches the target at low level before popping up at the last moment to acquire the target visually, the bombs thereby descending steeply while the aircraft is exposed to the enemy for the minimum amount of time. Where the enemy defences have been suppressed, as happened in the Gulf War, a medium-level dive attack can be used, again giving a good angle of penetration for the bombs. Laser-guided bombs (LGBs) were also dropped from medium level during the Gulf War, the targets being designated by other Tornados or by Buccaneers. The JP.233, however, did require the Tornado to fly at low level and fly directly over the airfield which the weapon was intended to destroy, and consequently a sizeable number of aircraft were damaged or destroyed during the early days of Operation Desert Storm because of the dangerous environment through which the crews were flying. The Allies quickly neutralised Iraq's air force, however, making further airfield attacks unnecessary. All these different types of attack are regularly practised during training sorties, with the crew flying either visual with the target or blind, because of weather or distance.

As the Tornado approaches to within two minutes' flying time of the target, the navigator will search for his pre-briefed radar fix points, which the pilot will also look for visually. The target can be marked by laser or radar, or visually through the HUD, and likewise, the bombs can be dropped automatically or manually. The Tornado's computer system is particularly accurate. For example, a typical laydown attack normally achieves a delivery within 10ft of the designated impact point. The JP.233, unique to the Tornado GR1, dispenses 215 area-denial mines and thirty cratering munitions, rendering the runway unusable and very difficult to repair.

'Bombs gone … recover' is the instruction from the navigator, signifying that the weapons have been delivered and enabling the pilot to rejoin the rest of the Tornado formation as planned, ready to resume the correct positions for defensive cross-cover. On peacetime training missions it is normal for the aircraft to attack two or three different simulated targets, some purely visually, or to use small practice bombs against weapons-range targets. This ensures that all aspects of weapon aiming, radar handling and weapons selection are practised again and again. Bombs dropped on the weapons ranges are scored relative to the precise target position, and the element of competition between crews provides an extra incentive to achieve the best results. The range of possible in-flight emergencies is quite large, although every system in the Tornado has a back-up or reversionary mode. Even the complete failure of one engine will not prevent the aircraft from flying safely. In peacetime, however, any major systems failure will normally require a sortie to be terminated. The aircraft will be flown to the nearest suitable and available airfield, and engineering advice will be sought. In wartime the mission would naturally have a higher priority, and the Tornado would be flown with a greater level of acceptable systems failures. In some cases, battle damage may require an aircraft to be coaxed home with severe and multiple problems, and the Tornado aircrews regularly practise such worst case situations in ground simulators. Coping with emergencies is a vital skill when one considers the number of things that could go wrong, ranging from major malfunctions such as hydraulic or electric failure, bird-strikes, a loss of aerodynamic control (stalling, spinning, etc.), burst tyres on take-off or landing and brake failures (leading to an arrester cable landing), to less catastrophic malfunctions such as instrument or radio failure. For every problem there exists a set procedure to assist the crew in achieving a safe solution. The flight crew check list (FCC), backed up with personal knowledge of the systems, will help the crew to solve the problem. When no other safe solution is available, however, the final option is the ejection scat. The Martin-Baker seat is a 'zero-zero' escape

ABOVE: An unusual view looking upwards at a Tornado in flight, illustrating the aircraft's weapon stations. The main weapons hard points are located directly under the aircraft's fuselage, and in this instance four 1,000lb inert practice bombs are being carried. (*Tim McLelland*)

ABOVE: Out over the North Sea, two Tornados from No 2 Squadron demonstrate the type's ability to launch flares, used in self-defence against heat-seeking missiles. The Tornado is expected to remain at the forefront of the RAF's offensive capability for a few more years until it is completely replaced by the Typhoon (in its swing-role attack version) and the F-35. (*Tim McLelland*)

system, able to operate at zero forward speed and zero altitude if necessary, thanks to an internal rocket pack which will fire the seat and occupant clear of the aircraft at a staggering initial acceleration of 25g! The seat includes stabilisation equipment, an oxygen supply and an automatic separation system, putting the occupant quickly under a fully deployed parachute. Attached to the pilot's harness

is a personal survival pack, with a dinghy and emergency rations, enabling the downed crew member to survive in a hostile sea (or land) environment until rescue arrives.

Recovering back to base, the formation will call Air Traffic Control three or four minutes before arrival to obtain the latest airfield status information. Instrument approaches are often necessary in typical

German weather, but otherwise, when conditions permit, a visual recovery is made, running in over the airfield, breaking individually down-wind, reducing speed and extending the wings to their fully forward position, with flaps, slats and undercarriage all deployed. The final approach is flown at about 135kts, and once the aircraft is safely down on the runway the nose can be held high to achieve some aerodynamic braking. The thrust reverser buckets can also be deployed when required, and for a full-performance, ultra-short landing the reversers can be pre-armed to deploy as soon as the main wheels make contact with the ground. When the aircraft is clear of the runway, one engine will be shut down and the weapon aiming equipment, some navigation equipment and the flight systems will also be switched off. The Ground Controller directs the crews back to their individual shelters, where the aircraft is winched backwards into the HAS and the engineers get to work on any problems reported by the aircrew. Back in the PBF the crews complete the usual supply of forms, detailing the route flown, the weapons dropped, the fuel used etc., before attending the debrief. The first topic of the debrief will be any flight safety points which may have arisen during the sortie, and any lessons to be learned will be stressed. The Flight Leader will then ascertain whether the object of the mission was achieved – whether each of the crews hit their targets and what the range scores were. Then each stage of the sortie will be discussed in detail, from brief through start-up, take-off and attack profiles, through to recovery and landing. After this, the radar and HUD film recordings will be assessed to verify the crews' claims regarding their results. This done, the mission is complete. From the beginning of the planning stage the whole exercise may have taken up to six hours, only two of which may have been spent in the air. The Tornado GR1 is a true fly-by-wire design, allowing the aircraft to be flown freely without any risk of accident. However, the Tornado systems do have limits, and the aircraft has to be kept within its design envelope. Even so, it handles well, and it is often referred to as a 'big Hawk' by its pilots. For its size and weight it possesses a good range and can carry a big payload, and compared with other bombers it has a proven record for the accuracy of its weapons delivery. Serviceability is excellent, thanks to the many LRUs (line replaceable units). The Tornado did of course receive a major avionics upgrade which equipped

it for continued operations, and in its current (and final) GR Mk 4 form, the magnificent Tornado remains very much in business, unless the governmental money-counters decide otherwise.

It was in Iraq that the Tornado earned its reputation as a formidable fighting machine. One experienced navigator provided the author with a fascinating first-hand account of what it was like to fly the Tornado in combat:

We'd always make sure that we were as full of fuel as possible before crossing the Iraqi border, at which stage the tanker would leave us and we'd press on in radio silence, although the AWACS E3 would be listening to us all the time. On night sorties there was always a fairly large number of aircraft crossing the border, and it always amazed me that nobody hit each other. Most of the aircraft by this time had switched their lights off but if you put on your night vision goggles it was like Piccadilly Circus – literally hundreds of aircraft going in and quite often you'd hear them checking in with AWACS before crossing the border. It was complete pandemonium. The run-in would start to get rather tense then, and we'd be busy double-checking all of our equipment. If any of our electronic warfare package wasn't working, that was the point at which you'd turn back, because you'd obviously need to know whether you were going to get shot down by something and you'd also need to know that your EW (Electronic Warfare) system is going to block any radar emissions which might be looking at you. So if that's not working you have to go home, which is something I had to do on one occasion. If you haven't been talking to your pilot before, this is the point at which you begin talking to him non-stop, as he will want to know everything you're doing in the back seat, and vice versa, so things get a little busy. Your eyes are on stalks, looking for anything you might see or hear, or see on the radar warning receiver. You're constantly talking to each other about how the timing is going, how the speed is, how you're feeling, and the conversation increases the closer you get to the target. Most of the targets that we went for in the Tornado were quite heavily defended, the airfields being the worst, but later in the war, even if we were going against the bridges, these too had tripleA (anti-aircraft artillery) sites close to them, as the Iraqis gradually realised which targets we were going for. Most of the airfields and POL (petrol, oil and lubricants) sites we had intelligence on, so we knew roughly what kind of defensive systems were around them, although sometimes we were surprised, finding completely different systems there. It was very comforting to be going in with American support beside you, so if any of the radar emitters on the ground did light up, then the Americans would have a go at them with their radar-seeking missiles. The Americans were very interesting to us. Their fighter boys weren't getting the number of kills they wanted, so quite often they'd be flying along beside us with all their lights on, flickering their afterburners on off so that any Iraqi fighter pilots would see them and hopefully have a go at them. But that didn't work – it just made us more nervous!

One of the most amazing things I've ever seen is tripleA coining up at me. Obviously I'd never seen it before, and to be suddenly looking straight at it was quite frightening. It just snaked up at you, and always seemed to reach the height we were flying at, so we'd be talking about it the whole way through. It just lights up the whole sky around you. Some of it comes straight up at you or around you and some snakes up towards you, and it was quite obvious that the guys on the ground were lying on their backs and were just waving the guns backwards and forwards in the hope of catching something. There was lots of flak around too, and you often felt it exploding around you as you went in towards the target. I didn't see many surface-to-air missiles, and many of them were unguided because as soon as their radar systems came on they'd be knocked out by the Americans. They tended to launch them blind most of the time, with no radar image, hoping that they'd hit something, and I saw a few of those come quite close to me. You don't see any sign of them on your RWR, but you'd suddenly see a missile come whistling over your canopy, fully armed and ready to go bang if it hit you. As for evasion, well you'd try to go round a tripleA battery the best way you could if you saw it, and likewise, if you saw a missile you'd pull hard to avoid it if at all possible. You talk to the pilot all the time and he talks to you, because if he's going to pull the aircraft violently to one side you obviously want to know what he's doing. Contrary to

popular belief, the Iraqi fighters were airborne in the early stages of the war. There weren't all that many, but they were definitely there and I was locked on by something on the third or fourth night … You do what you can to avoid it and you drop chaff out the back or you fly some fairly violent manoeuvres … whatever you can to break that fighter's lock on you, otherwise he's going to launch a missile at you. As we closed in on the target the pilot would basically want to know how far away we were and how the timing was going. The level of conversation goes up one hundred per cent. You're talking non-stop. The radar was switched on at quite a late stage, so as not to warn the Iraqis that we were coming in, and you only used it for short bursts at any one time, to identify the radar offsets and eventually to identify the target. You'd be looking at fixed points all the way along the route from Bahrain, but only for very short periods, so that the radar is never on for very long. However, you have to make sure that your navigation equipment is very good, so that by the time you get to the target you don't have any problem finding it on radar.

The radar is very good and you'd get an excellent picture in the back cockpit of the target area. As long as you've done your route study correctly, you'll have no problem finding your target and the exact point that you want to hit. The pilot is more concerned with looking outside the aircraft, watching for other aircraft, missiles or tripleA, while the navigator is more concerned with making sure that the bombs go off on target and that the timing and navigation are working okay. You make sure that the bombs are live. It's a two-man aircraft, so both people have some means of controlling bomb release, and you have to make sure that you're both happy before they're dropped. You go through all your switches, making sure everything is live and ready to drop on command. It's all computer-controlled, and all the navigator can do is to tell the computer where to drop the bombs. It will then work out the trajectory where the bombs will go, and it will drop the bombs automatically when you reach the release point. The bombs will fall off, and you'll hear them drop in sequence … pop … pop … pop … as they come off, and you count them to make sure they do all drop, or maybe you'll make a quick jinking manoeuvre to actually see them going down to the target. You then turn and head for home as fast as you possibly can. The pilot is obviously concerned that you're laser marking the right place, and he'll be continually talking about this. He'll want to know that all the bombs are off and that the switches are all looking good. The aircraft is much lighter now, much more manoeuvrable, as you've just got rid of quite a few thousand pounds. If you do feel under threat at all, you can plug in the afterburners to give you a bit of extra speed in the turn to get you back home again. It's basically a race to get back over the border and rendezvous with the tanker, although quite often you didn't even need one. But he'd always be there, and wait until the last man was over the border. If one of the boys didn't come back, the tanker captain would wait until there was absolutely no hope at all of the guy coming back. The tankers guys were excellent. It's the navigator's job to guide the pilot into the refuelling basket, and if the pilot was feeling a little shaky after flying around Iraq it could take quite a while to get back into the basket. You always took just the minimum amount of fuel you needed, as you didn't want to waste time ... you wanted to get back on the ground to have a good rest, so you'd unplug from the tanker and get out of the way, ready for another receiver who might be shorter on fuel than you were. Once when we arrived back at Bahrain we were just listening to the controller giving us clearance to land when we heard the air raid siren go off in the background, meaning there was a Scud missile inbound to Bahrain. We were completely exhausted but we had to hold overhead the airfield and plug into a tanker, staying up there for nearly half an hour.

How did the trip to the Gulf go? Well, I guess it went well because I came home and that's the main thing. The big concern is not getting shot down, but it's a worry, making sure that you do your job right and that you don't let down your mates. You're nervous when you go out to the aircraft and when you get back, but as soon as you get into that aircraft you've got a job to do and you have to put the nerves behind you. You have to make sure you do the job properly. The transit to the target, probably from the point you take off to the rendezvous with the tanker on the outbound leg is the time when you can think about what you're doing and you can start to get nervous. When you're actually tanking you have a job to do so you're okay. You get very nervous indeed when you leave the tanker,

before crossing the Iraqi border, but after that you have so much to concentrate on, you just haven't got time to be nervous. On the target run you're just too busy, and you just have to make sure you get those bombs on the target and destroy it. Coming home, you just want to get back as quickly as you can, and as soon as you step out on the ground you feel nervous again, thinking my God, what the hell have I just been through? Very mixed emotions, really. Obviously one or two missions were particularly scary, and when we lost people it was particularly disappointing, but it's one of those things that you have to get over. You've got to go and fly the next day, probably to attack the same target, so if you do feel upset because you've lost a colleague you have to put it behind you and get on with it. Sounds very cold, but you can't let things like that affect you, otherwise you might be the next one to go down. Sortie lengths varied from maybe three and a half hours to five and a half hours, the longest I flew being well to the north of Baghdad. I was feeling pretty shabby after that one.

Back home I'm now an instructor, and it's my job to train new navigators to go and fly in the Tornado. Because of what I did in the Gulf I can train them with an emphasis on actually fighting a war ... when I was a student the last thing on my mind was fighting a war. However, when it did happen I was ready and prepared for it and I like to think that I'm training my students with war in mind. I tell them Gulf stories, and if they do make mistakes I liken it to a war situation, which makes for a good learning environment. In a strange way the Gulf War was very satisfying as it was a chance to put into practice what I'd only done in theory before. You can always train for war, but you can never know how you'll really perform unless you've actually done it. Now that I have, and flown eighteen missions, I figure I must have done a good job to survive it, as did my colleagues. To have flown inside enemy territory is a very frightening experience, but to have come back from it, having achieved my job of putting bombs on the target, makes me feel very good. It makes me feel proud of both myself and the rest of my colleagues too. I'm far more confident now, and it's also made me appreciate the fact that life can stop at any moment, which makes you want to live life to the full. Certainly that's something I've done since I came back, but maybe not something I'd done before the war. While I was actually out there in the war I realised that at any second I might no longer exist. So now I want to make sure that I live my life to the full.

Another interesting point, and a nice one for us, was the amount of media interest. The press were very reasonable to us and we had all kinds of people writing to us, ranging from mothers and grandmothers to schoolchildren. We still correspond with some people now. We got letters and good-luck cards from lots of people, even chocolate bars and teddy bears, which we took flying with us. Valentine's Day was fantastic! Much of the combat flying was essentially similar to peacetime flying. The planning stage was much the same, as we always simulated an EW environment, we often flew with package support and we dropped bombs on a range. The only real difference was that whereas much of the peacetime flying is good fun, the Gulf flying was taken much more seriously. But we'd never underestimate the value of training. Much of the specific planning was the same as during peacetime, the pilots choosing the actual places at the target to attack and choosing the offset points and so on. The navigators still did the route planning in and out of the target area. Perhaps the real peacetime difference is that you're not in danger from a Scud missile coming through the roof while you're planning or briefing, so you wouldn't be carrying your Nuclear, Biological and Chemical warfare suit around with you. However, no matter what training you do, you can never be completely prepared for the real thing. We always thought that if we did go to war it would be an East–West conflict, and we certainly didn't think we'd be flying in the Middle East. So to go and fight in the desert and have camels to avoid, was very different! Above all, we proved that the Tornado was more than capable of fighting in a modern environment. That's more than can be said for some modern aircraft.

Like the Tornado, the Jaguar was also a vital part of the RAF's inventory through the 1980s and 1990s. Oddly, the Jaguar was initially designed as an advanced trainer, but the Anglo-French aircraft eventually emerged as a strike/attack aircraft, suited to the European theatre of operations. It's fair to say that the RAF didn't expect a great deal of it when the Jaguar first entered service (it was regarded as a child of politics) but its versatility, performance and reliability soon changed that perception and by the time that the Jaguar was withdrawn from service it was regarded as an outstanding machine, and had it not been for a growing need to reduce defence expenditure, the Jaguar might well have remained in RAF service for many more years. No 54 Squadron was one of three Jaguar units stationed at Coltishall in Norfolk. It was allocated to NATO's ACE Mobile Force, and its primary role was to support, from bases in Denmark, NATO's Northern Flank in the event of a war with the now-defunct Warsaw Pact. However, following the break-up of the Soviet Union, the RAF's Jaguars were employed in different roles, not least during Operation Desert Storm, when they were operated very successfully, without incurring a single loss. Jaguars remained in the Middle East, a presence being maintained in Turkey for some time. Low running costs, ease of deployment and high reliability were all factors which favoured the Jaguar, and contrary to earlier plans for its swift replacement by the European Fighter Aircraft (Typhoon), it remained in RAF service for a long time, following a decision to first re-equip the RAF's Tornado F3 interceptor squadrons with Typhoons. Coltishall was home to two other Jaguar squadrons – Nos 6 and 41 – while No 16(R) Squadron at Lossiemouth acted as the Jaguar OCU. Each squadron was equipped largely with single-seat Jaguar GR1As (later upgraded to Mk 3 standard), although small numbers of two-seat T2As were also operated (the majority with the OCU), for check-rides, dual control instruction, etc. The GR1A was primarily a ground-attack aircraft, although some export models of the Jaguar are employed as air defence fighters, and some of the RAF Jaguar fleet assigned to Germany were also capable of operating in the strike role, armed with the WE.177 nuclear store.

Nominally, approximately twelve aircraft were assigned to No 54 Squadron, together with sixteen pilots. The number and type of missions flown each week depended largely upon aircraft serviceability and whether the squadron was working up to any major exercises. For example, a pilot could fly perhaps three sorties in one day and none the next day. However, on average, a pilot could expect to fly thirty hours a month. A typical mission would last for roughly ninety minutes, unless the sortie included air-to-air combat, in which case the

ABOVE: A gaggle of Jaguars make their return to RAF Coningsby after completing a formation flight prior to the type's retirement from RAF service. Clearly visible here are the Jaguars' huge air brake doors, positioned behind the aircraft's main landing gear legs. These (together with a brake parachute) gave the Jaguar a very impressive short field landing capability. (*Tim McLelland*)

duration dropped significantly to around forty minutes because of the thirsty characteristics of the Jaguar's Rolls-Royce Adour engines when reheat is selected. As a writer, I gratefully accepted two opportunities to fly in the Jaguar, the following being an account of a typical mission conducted by No 54 Squadron from Coltishall:

The tasking for each mission is normally received two hours before the scheduled take-off time, the standard NATO timescale. The briefing begins with a time check, followed by the aims of the sortie, call signs, a listing of the pilots, the aircraft serials, the weapons fit each aircraft is carrying and the aerodynamic limits to which each can be flown. Spare aircraft are detailed, together with the weather for the local and target areas, the amount of fuel needed to return to base, and who will be responsible for making the various fuel calls and radio frequencies, which are pre-briefed to avoid the need to broadcast such details when in the air. The Jaguar is equipped with 'Have Quick' frequency-agile radios, which are virtually resistant to jamming. The briefing continues with details of the take-off, recovery, routes to be flown, heights, Notams, Royal flights, safety altitudes, emergencies, air-to-air refuelling (if it is to be employed) and more. The evasion brief will detail the types of aircraft which the Jaguar pilots will be avoiding, the limits to which evasion will be flown, and the tactics which are to be employed. The whole briefing will last up to forty minutes, depending on the experience of the pilots. If the formation includes a 'new boy' the brief will have to include precise details of what is going on so that he is left in absolutely no doubt as to what he will be expected to do.

The planning for each mission is time-consuming, but it is vital to the success of each sortie. The Jaguar pilots adhere to a much-used acronym – KISS (keep it simple and safe). The target is first examined and weapons are then assigned as appropriate, after which the ground crews will be advised of the weapons fit required for each aircraft. The route to the target is studied, threats are taken into account (with a great deal of input from an Army ground liaison officer assigned to the Squadron) and the delivery techniques (level bombing, dive bombing, or other tactics) are chosen. The Jaguar is equipped with a FIN.1064 inertial navigation platform, a very

accurate system that runs a projected map display and generates steering information in the head up display. If times for waypoints are programmed into the computer, it will also give demanded ground speed and actual ground speed, enabling the pilot to arrive at his designated target precisely on time, provided that he adheres strictly to the HUD information. TOT (time on target) is vitally important so as to avoid conflict with other strikes, reconnaissance overflights or friendly manoeuvres on the ground.

The INS requires specific target details, such as its latitude and longitude, its height, so that the HUD can generate a bar over the predicted target position before it becomes visible to the pilot, and the initial point (IP) position, so that any wandering by the INS can be updated between thirty and ninety seconds before the aircraft arrives at the target. Some pre-IP updates can also be fed into the INS, the most accurate means of revising the INS's data being to use the Jaguar's laser target seeker head, although this is strictly a wartime option as the Jaguar's laser is not eye-safe. Indeed, the laser can be used in peacetime only over weapons ranges, and only within a few hundred yards of the target position. The IP and target coordinates become two of a series of waypoints that form the complete route that is fed into the INS.

The RAF Jaguar fleet utilises a Total Avionics Briefing System (TABS), which enables the pilot to feed all the relevant data into the aircraft swiftly and easily. He lays his map on a digitising map table and programmes each waypoint into a data store by means of a small hand-held cursor. A TV monitor displays the information as it is programmed, enabling errors to be corrected as required. Once the route is properly programmed, the 32k memory portable data store (PODS) is ready to be plugged into the aircraft. A record of the flight is given to an operations officer, who will book the aircraft into the low-flying system and also into whatever weapons ranges are required. Departure out of the airfield is also booked with Air Traffic Control so that the Jaguars can be handed over to a radar control authority if required, although for many missions the aircraft will depart visually at low level. Part of this planning must include careful timing of the attack phase, so that a first-run attack on a weapons range is timed to coincide with the operating periods at the range (which closes at various times to allow, for example, targets to be changed). Therefore TOT and time on range must match. A formal flight plan is filed only if the aircraft are expected to enter the airways network or if they are flying overseas.

When planning has been completed it is time to gather the appropriate flying kit together, which includes the bulky immersion suit if sea temperatures are below 10°C. An anti-g suit is worn, together with the usual gloves, life preserver and flying helmet. Once the pilot has been fully kitted out, it is time to walk, the next stop being the Operations desk, where a final out-brief is given by the authorising officer. The crews then walk to the line hut to sign out each aeroplane (on RAF Form 700) checking that each is fuelled correctly, is carrying the required weapons fit and is fully serviceable. Unacceptable defects are always noted in red ('I always look for missing wings or wheels,' commented one pilot) and acceptable defects are also noted, these being regarded as relatively unimportant provided the pilot is aware of them. Individual aircraft also develop specific problems that tend to recur, and each pilot is therefore careful to note any previous defects in case they arise again during the flight.

Arriving at his aircraft, the pilot makes a preliminary inspection, ensuring that it is parked correctly, that chocks are fitted under the wheels, that safety pins are fitted to each weapon, that external equipment is available and that a fire extinguisher is present. He then climbs into the aircraft, checking that the cockpit's circuit-breakers are all set and that the battery, ignition and parking brake are all on. The ejection seat pins are confirmed as being fitted, and then the rudder pedals and the seat are adjusted as required. Looking to the left side of the cockpit, the pilot checks that the undercarriage handle is down, that the canopy jettison handle is flush and that the arrester hook handle is fully forward. Behind each throttle is an igniter re-light button, and a clicking sound will confirm that this is functioning correctly. The master armament safety bus bar key is then fitted into its slot, and the pilot climbs out of the aircraft again to complete a full external check. The first task is to look into the nose wheel bay, where he selects a number of switches to indicate the weapons fit and the ballistic mode in which the weapons will be dropped. The AN/ALE40 chaff and flare dispenser selections are

also made in the nose wheel bay. The checks then continue with a look at the various vents, to ensure that they are all unblocked. The hydraulic accumulators are checked for the correct pressure, and the safety pins are confirmed as being removed. On a more general note, the pilot will look for any leaks and, if necessary, check with an engineer that they are within limits. He will also look for any loose panels and unusual cracks, check inside the engine intakes for foreign objects, check the exhausts, confirm that the afterburner rings are in good condition and also ensure that the fuel shield inside the engine bay is not likely to fall off. The removal of the arrester hook pins and brake parachute pins is also checked.

The Jaguar is fitted with an internal starter unit, which incorporates a microswitch that needs to be placed in the correct position, and the starter's oil level is checked at the same time. It is also customary to see pilots shaking the various external stores on their pylons, ensuring that they are all properly secured. Once satisfied that the aircraft is in good condition, the pilot can climb back into his cockpit to complete the pre-start checks. Strapping in is the first task, working from left to right, attaching the personal equipment connector (oxygen, anti-g air pressure), personal survival pack (dinghy, etc.), leg restraints, lap and shoulder straps, RT lead and helmet and removing the seat safety pins, at which stage the 'liney' (the member of the ground crew who assists with strapping-in) will remove the access ladder. Working from left to right again, almost every item of equipment in the cockpit is checked, starting with the wander lamps and auto-stabilisation systems (pitch, yaw and roll) and also checking that the laser is off. The flaps are moved to 'up' although without hydraulic pressure nothing happens at this stage. The inertial navigation system is normally switched on prior to the pilot's climbing in, giving the equipment time to heat up and align itself with its position. The pilot plugs in his data pod and selects 'DTS' (data transfer), checking that the route is programmed into the computer properly. Throttles are checked for full and free movement, and the main flight instruments are all scrutinised to ensure that they are functioning properly. The HUD is turned to the correct mode, usually 'rad alt' (radio altimeter) for take-off. On the right-hand side of the cockpit are various fuel gauges and warning panels; the fuel cross-feeds are checked, as is the EHP (electro-hydraulic pump) which supplies hydraulic pressure (if both pumps fail in flight). The engine instruments are examined, and then the alternators and transformer/rectification units are switched on. The air-conditioning is switched to 'ground' and the Tacan (Tactical air navigation equipment) is switched on. It is now time to start the engines.

The pilot makes a hand signal to the ground crew and the micro-turbo starter is switched on, taking four or five seconds to run up to 85 per cent. Once the unit is at this idle setting, the pilot holds one finger in the air, vertically, to signify engine No 1 start. He opens the low-pressure cock and presses the start switch. When the engine has wound up to around 52 per cent, the pilot checks that the relevant captions are out and that the flaps are being raised to their selected positions. No 2 engine is then started, and the external power is switched off. Each engine incorporates a dump valve that releases excess air from the compressor during start-up, and advancing the throttles to 61 per cent will close the bleed valves, confirmed by a slight reduction in the TGT (turbine gas temperature). The engines idle at between 54 and 57rpm on the ground. The flying controls are checked, and the auto-stabilisation system is checked. Flaps can be moved to one of eight settings, half-flap being 20 degrees and full flap 40 degrees. The flaps and leading-edge slats run along the entire leading and trailing edges of the wing, turning being achieved by the use of spoiler devices. For take-off, flaps are set at 20 degrees.

Radio check-in with the rest of the formation begins at a pre-briefed time, the pilots using twenty pre-set frequencies, Stud One being for 'ground' (start-up and taxi). The INS is switched from its alignment mode, and with nose wheel steering selected (by means of a switch on the control column, which also operates cameras in the Jaguar reconnaissance model) the aircraft is ready to taxi. With roughly 70 per cent power on each engine, the parking brake is released and the aircraft rolls forward, at which stage the pilot tests the foot brakes and returns the throttles to idle before turning left or right so as to avoid damaging other aircraft (or personnel) with engine blast. While the pilot is taxiing to the active runway there is time to consider emergency procedures for take-off. If an engine fails

on the run below 100 knots, the take-off can be aborted by bringing the throttles back to idle and streaming the brake parachute; the aircraft will be very heavy, putting a great deal of strain on the undercarriage. Above 100 knots the one-shot arrester hook can be extended (it has to be raised by an engineer). Alternatively, a single-engine take-off can be made by raising the undercarriage and dumping all external stores (by means of a single emergency switch). However, the pilot must be careful not to dump weapons on top of the travelling undercarriage. Speed is life as far as Jaguar pilots are concerned – the faster the aircraft flies, the lower the angle of attack (AOA), and the primary aim on a single-engine take-off is therefore to achieve speed, not altitude.

At the end of the runway, the INS clock is started and the radio frequency is changed to 'Tower' enabling the pilot to obtain departure instructions. The aircraft then lines up on the runway, perhaps with others. Four aircraft is the largest number which can be accommodated line-abreast on the runway and, once in position, each pilot checks the runway caravan for any safety signals, checks that all warning captions are out and checks that the INS clock is running. The throttles are then pushed to full dry power while the aircraft is held on the foot brakes and the EGT is within its limit of 700°C. With a thumbs-up from the wingman, the leader will make a chopping motion with his hand, to signal brake release, and the aircraft begin their take-off rolls. Aircraft depart solo, in pairs or in pairs of pairs, separated by thirty seconds (equivalent to about 3 miles).

After another nod from the leader as his aircraft passes about 40 knots, the pilots release catches on the throttles to select afterburners, checking that flow of fuel increases (although the Adour engine is very reliable, and faults rarely occur). At the pre-briefed rotation speed (dependent upon weight and temperature, although it is normally around 180 knots heavyweight and 140 knots lightweight) the aircraft is rotated to 14 degrees AOA and the Jaguar is swiftly airborne, normally at around 5,000ft from brakes release. Once airborne, the aircraft quickly move into formation, achieved by lining up wing and tail aerial positions visually. The undercarriage gear and flaps are retracted, and the

AOA audio warning and the air-conditioning are switched on (air-conditioning drains 4 or 5 per cent of thrust from the engines, so it is not switched on until the aircraft is safely airborne). The Jaguar is a very sprightly performer when flown 'clean', requiring only 2,000ft to take off, but most missions are flown with a pair of 924kg (2,035lb) external fuel tanks under each wing. Moreover, in hot weather the Adours do not perform very well, and in extreme conditions of perhaps 30°C, with a full fuel and weapons load the Jaguar GR1 would require a runway of 7,000ft or more. However, the engines can be modified to give additional thrust, as was the case during Operation Desert Storm.

Once airborne, the pilot will ensure that the Jaguar's defensive systems are functioning, these being the ECM (electronic countermeasures), RWR (radar warning receivers) and chaff/flare dispensers. The formation will then adopt tactics which are appropriate to either a ground or an air threat as necessary. The Jaguar is fitted with a radar altimeter, which is considered to be very accurate (pilots confirm that it reads just 5ft altitude whilst taxiing), and this enables the aircraft to be flown at ultra-low altitudes … a mere 15ft above the sea is quite practical, the only real limits being the risk of scraping external stores and of course the sheer nerve of the pilot. However, such low flying is an exception rather than the rule, and in the United Kingdom the normal 250ft minimum applies and only occasionally is special permission granted for 100ft AGL flights in strictly designated areas. Jaguars normally fly in battle formation, in line-abreast with 2–3,000yd of separation, ensuring that a hostile aircraft cannot close in on the tails of the aircraft. The entire formation can fly in line-abreast or in 'card' formation, in trailing pairs. Low flying is still considered to be a key to survival in a hostile environment, even though RAF Jaguars operated at medium altitudes over Iraq and Kuwait. The Allied air forces were able to maintain a sterile air environment in the Gulf War, allowing attack aircraft to operate at altitudes which would normally invite surface-to-air missiles and anti-aircraft artillery. Obviously the same kind of relatively benign operating conditions may not be available in any future conflict in which the Jaguar might participate. The high wing-loading characteristics of the Jaguar make the aircraft a good performer at low level, offering a safe, smooth and quiet ride, with no smoke trail and only a small radar return. Indeed, during severe contour flying, the Jaguar can hug the landscape, the pilot rolling the aircraft through 180 degrees to make a positiveg pull downhill rather than the usual negativeg bunt. However, such manoeuvring tends to be reserved for lightly loaded aircraft, as a full weapons load will significantly reduce the aircraft's roll rate.

As the aircraft reaches each turning point, the HUD changes to a loose navigation mode and a marker appears over the predicted position of the waypoint. As the aircraft flies over the waypoint, the pilot will update the INS position if necessary and select the 'change destination' button, switching the HUD back to navigation mode. When the aircraft approaches the initial point, the INS will display a cue in the HUD, sixty seconds from the IP. The pilot will check his bomb selection, and check that the pylons are 'on' and the weapons fused, check the 'stick spacing' (the interval between each bomb drop) and select auto attack by computer or manual attack and guns 'on' or 'off'. Then the late arm safety switch is selected, making the whole system live. Finally, a small catch on top of the control column is activated, enabling the computer to drop the bombs at the required time. Targets are usually acquired visually, although the laser can be used, giving an appropriate cue on the HUD. For a typical automatic attack, the pilot places a target bar over the target position, using a small controller located behind the throttles. This ground-stabilised image is automatically connected to the laser ranger, which fires at the target and measures the range by calculating the time taken for reflected laser light to return to the aircraft. Keeping the fire committal button pressed confirms the attack, and when the CCIP (continuously computed impact point) and target become coincident, the weapons are released automatically. The attack sequence takes place in a matter of seconds, as the Jaguar pilots will try to unmask (reveal themselves to the target and its associated defences) at the last possible moment thus leaving themselves very little time to acquire the target. Terrain-masking is still recognised as being tactically valuable, and the Jaguar pilot will try to dash in and out of the target area as rapidly as possible before disappearing in the surrounding 'clutter' again.

ABOVE: Jaguar XX725 first flew in 1974 and spent some time serving with the Indian Air Force before returning to Britain to join the RAF. After retirement, it was transferred to RAF Cosford where it continues to be used as a ground-based systems and maintenance training airframe. (*Tim McLelland*)

On departure from the target area, reheat can be used to pick up extra speed if necessary, although, as one pilot commented, 'If you select afterburners, every heat-seeker in the vicinity will he going "Yum yum, there's a heat source" so you have to be very careful.' The Adour engines are not immensely powerful, so they are not particularly hot either, and they give the Jaguar a fairly small heat signature, especially when compared to aircraft such as the Tornado. Once back over friendly territory, the formation can raise its altitude and assume a relaxed non-aggressive flight attitude so as to avoid being shot down by friendly forces. IFF is a vital piece of equipment for sorting the good guys from the bad guys. The flight back to base is normally made at around 360 knots as this is a good speed for low fuel consumption in most weapon configurations. On a typical peacetime return to Coltishall, the formation will make radio contact with Squadron Operations, giving prior notice of its return to the waiting ground engineers. Selecting 'Stud Three' puts the formation back into contact with Air Traffic Control, and a visual recovery is normally made. The final arrival over the airfield is made on the control tower radio frequency, the aircraft flying in a loose arrow formation. Thirty seconds before arrival overhead, the call 'Initials' is made and the aircraft break left at 1,200ft, in a 4–5g turn, air brakes deploying and throttles moving back to idle. The aircraft lose speed rapidly, and by the time they are established on the downwind leg of the circuit it has been reduced to 260 knots. Flaps are now extended 40 degrees, air brakes are in and speed is down to 230 knots. Gear down selection can now take place, and a radio call is made to state intentions, either to land or overshoot. The harness is checked tight and locked. Fuel is checked to be sufficient, with hydraulics working, three greens (undercarriage down and locked) and rudder sensitivity changed from small to large, and full flap is extended before a 12-degree AOA turn is made on to final approach. Engines are set at roughly 93 per cent, and the turn will be tightened or widened to take the direction and speed of the wind into account. The 12-degree AOA is maintained, and a careful check is made that the engines are functioning correctly … if one fails at this stage, the pilot is obliged to eject. Once the aircraft is established on finals at about a mile from touchdown, the gear is confirmed as down again

and the aircraft is positioned on the correct glide path to the runway, usually relying on PAPIs (precision approach path indicators), which should read as two reds and two white lights, changing to three reds and one white just prior to touchdown. The aircraft is aimed to land 'on the numbers' just beyond the 'piano key' threshold markings at a touchdown speed of around 140 knots. Once it is down, the throttles are closed and the aircraft is flared by rotating upwards to 14 or 15 degrees alpha (AOA) – anything greater would probably cause the under fuselage strakes to make contact with the runway. The nose will automatically fall at around 120 knots, and nose wheel steering is engaged, with braking commencing at below 100 knots. A heavy landing or an unusually fast landing can be retarded by releasing the brake parachute, the deployment handle for this being located on the left-hand side of the cockpit. Once the aircraft is clear of the runway, one engine is shut down. A lightweight Jaguar can tend to run away with its occupant; with both engines running at idle this is not a particular problem, but it can cause the brakes to overheat. Armament switches are 'safed', ejection pins are replaced, the canopy is partially opened and,

ABOVE: A Jaguar-pilot's view, high above the North Sea, looking out across the aircraft's square-shaped air intake and port wing, as three aircraft from No 6 Squadron close in to head home to Coningsby. (*Tim McLelland*)

once the ground marshal is spotted, the taxi lights are switched off prompting the marshal to guide the pilot back into the parking position. A final INS fix is taken, to establish the system's accuracy, and the engine is shut down. After signing for the aircraft, giving details of the INS performance, sortie, fatigue counts, etc., the pilots return to the squadron building to debrief. This establishes whether the mission achieved its aims and what may be learned from it. The formation leader will conduct the debrief, the supervising officer raising appropriate points as necessary.

The Jaguar does possess one or two vices, not least the T2's inability to recover from a spin. The aircraft can assume a dangerous flat spin from which a pilot is unable to eject, and it is therefore important to recover or get out of the aircraft quickly. The Jaguar's overall speed is good, and with afterburners the Jaguar can achieve Mach One at low level. Maintenance demands are very light; the Jaguar force can be deployed and return home without any major unserviceabilities, an achievement that can be matched by few other aircraft. The weapons capability is good, the RAF force carrying 1,000lb bombs with retarded or slick tails and air burst, impact or delay fuses, laser-guided bombs, CRV7 air-to-ground rockets (Mach Four speed, 10lb warhead), two ADEN 30mm guns, CBU87 or BL.755 cluster bombs, ALQ101 ECM pods and chaff/flare dispensers. Performance is good, especially when flown clean ('basically a Hawk with afterburners'), although the aircraft normally carries a hefty weapons and fuel load. It can also carry over-wing AIM9L Sidewinders, a capability developed for operations in the Gulf, and although they rely on bore sight positioning they do enable the aircraft to defend itself, if necessary.

Before it was finally retired, the Jaguar was upgraded and the later versions of the aircraft had better avionics and more powerful engines, making the Jaguar an even more formidable machine. As a writer (and a very lucky passenger), I couldn't fail to be impressed by the Jaguar. It looks bigger than it actually is (largely because of its stalky undercarriage, designed to give the aircraft a useful rough field capability), and once in the air it is therefore surprising just how fast and agile it is. The occupant of the rear seat in the trainer variant enjoys a magnificent view both to the left and right of the aircraft and directly over the forward occupant's seat, and as we battled through the cloud and rain across Cumbria during one dark afternoon, I could easily convince myself that I was alone in the aircraft, staring straight ahead through the glowing HUD. My most memorable experience of the Jaguar, however, was when I found myself flying along the downwind leg of a landing at RAF Coltishall, and suddenly received an invitation from my pilot to perform the landing myself. Naturally, I accepted the challenge, but hearing the words 'You have control' is quite unnerving if you don't happen to be a pilot. But much to my relief (and delight) the Jaguar is a fairly docile machine, provided that it is treated with respect. It took very little effort to gently bring the aircraft round onto final approach, and with some judicious verbal guidance from my pilot, I carefully manoeuvred the big cat onto a fairly respectable landing approach, eventually kissing the runway with our main wheel tyres not too far from the landing threshold.

8

AIR COMBAT

As a child of the 1960s, and having been obsessed with aeroplanes for as long as I can remember, it's fair to say that I grew up with the Lightning. My first memories of aircraft were forged at the Battle of Britain 'At Home' days held at my local RAF station, Finningley, the annual open day being the highlight of my year when it was possible to spend a whole day just absorbing the sights, sounds and smells of aircraft. It was a heady mix of excitement, noise, speed, spectacle, kerosene and hot dogs, which held me spellbound. The aerobatics were fine, the Spitfires and Hurricanes were interesting, but when you're just a kid it's all about noise, so the traditional Vulcan scramble was always the highlight of the show, just for the sheer thrill of feeling the ground shake as the dark, camouflaged monsters roared by. But there was always another highlight too – the sleek, shiny and almost brutal-looking jet fighter that was usually tucked away in a far corner of the airfield where the flying display aircraft rested between performances. When you saw the big silver tail fin slowly moving by above the crowds of spectators it was time to grab a good vantage point for the star performer. You knew it was coming though, as the first task of the show day was to check the display programme for the bits that you didn't want to miss. In those days there wasn't much of a list to read, as the flying display generally lasted only two or three hours at the most, so it took only a few seconds to thumb down the paper to find what you were really looking for – the Lightning. Of course, it never failed to impress. Glistening silver metalwork, big bright roundels and a splash of heraldry on the tail, two big and very serious-looking missiles strapped to its sides, swept-back wings, and a fierce nose complete with a long, pointed probe at the tip. It looked every inch a fighting machine. Once into the air, the Lightning more than lived up to a child's expectations. It seemed to perform effortlessly through loops, rolls, slow passes with its clumsy-looking landing gear extended, and low, fast passes accompanied by a plume of reheat flame, a terrifying cloud of water vapour and a gratifying, ear-splitting blast of noise which, as the great Sir Douglas Bader once said, almost encouraged you to cover your ears and stick your head under the nearest seat. But for a kid who was obsessed with aeroplanes it was different – I just wanted to stand there and enjoy the thrill, my senses battered by the noise, and when the Lightning made the final near-supersonic rush into a final vertical departure I was truly spellbound, watching as the great machine thundered higher and higher until it disappeared into the glare of the sky. Things just didn't get any better than that. Back in those days I never imagined even for a minute that one day I would be able to climb into a Lightning and go fly. Why would I have? Lightnings were almost mythical creatures that performed their mysterious business tantalisingly out of reach beyond crowd barriers and perimeter fences. They were strange and beautiful machines, handled by a special breed of men – superheroes dressed in their bizarre space suits. And yet, a decade or so later, there I was, high on top of the Lincolnshire Wolds, standing face to face with the magical monster – and it was inviting me to come inside.

LEFT: Lightning F Mk 2 XN728 wearing the markings and blue trim of No 92 Squadron at Yeovilton in 1965. The unit was then based at RAF Leconfield, but moved to Gütersloh in Germany shortly thereafter. (*Tim McLelland collection*)

Of course, I count myself as being very lucky. Not only for the opportunity but for getting there just in time. It was 1987, and the Lightning was on the verge of retirement from RAF service. The days of pristine, gleaming silver jets had long gone, and in front of me was something which seemed so very different from the aeroplane I remembered from my childhood. It didn't glitter at all – it was just plain dirty. The shiny metalwork was all gone, replaced by a dark and gloomy mix of green and grey camouflage which had obviously been applied to the aircraft for many years, the mismatch of camouflage lines revealing where access panels had been continually removed and replaced, some from different aircraft with slightly different colours, and most of them covered with scratches and streaks of dirt, their edges worn and chipped. Patches of fuel stains and oil seemed to cover every surface and under its belly drips of oil and fluid pooled onto the stained concrete, which had supported the weighty presence of these aircraft every day for a quarter of a century. Up close, the Lightning doesn't look quite so sleek; in fact the most striking impression is one of sheer size and bulk, the cockpit towering high above your head atop a clumsy-looking ladder which is clipped to the fuselage side. It looks big and brutal, but by walking round to the rear of the aircraft, you're left in no doubt that this machine is built for speed. The two soot-covered jet exhausts are huge and cavernous, big enough to climb into, and it's difficult to even imagine the endless torrent of burning kerosene, bursting through the ring of mechanical reheat petals and out into a long sheet of white-hot flame. But you only need to look into the mouths of the monstrous Avons to know that they mean business.

Naturally, getting airborne in a Lightning takes some preparation. Before I had even got anywhere near Binbrook I was required to spend a long day at North Luffenham, a former Canadian Sabre base, where an RAF aeromedical expert delighted in telling me and the assembled class of students about the varied and exotic ways

ABOVE: Lightning F Mk 3 XP696 was a well-known aircraft on the UK air display circuit for some time, performing dazzling (and noisy) aerobatics for spellbound audiences across the country. It is seen here racing skywards above the Gnat trainers of the RAF's 'Red Arrows' aerobatic team. (*Tim McLelland collection*)

in which we could get ourselves killed. Fast jets are not sports cars, and they have to be treated with the greatest of respect, especially when they routinely operate in an environment that is completely hostile to human existence. As if to emphasise this point, our class was treated to a spell inside Luffenham's decompression chamber so that we could experience for ourselves what would happen if the Lightning's pressurisation failed at altitude. Finally, it's time to get acquainted with my pilot (Flt Lt Mark Ims from No 11 Squadron) and our aircraft, Lightning T Mk 5 XS458, a long-time Binbrook resident which had served with both No 5 Squadron and No 11 Squadron prior to joining the Lightning Training Flight, whose blue lion motif was proudly displayed on the tail fin surfaces. First flown in December 1965 (little more than three years after I was born), XS458 first served with No 226 OCU at Coltishall although as a relatively late arrival, it did not get to wear the flamboyant red, white and dayglow orange markings which adorned the first T5's assigned to the OCU. Nearly twenty-two years later, XS458 was still very much in business as the LTF approached their last few days of activity prior to disbandment.

While Mark conducts a careful inspection of the aircraft's exterior, I carefully make my way up the aircraft's access ladder and – at what already seems like an indecently high altitude – gingerly insert myself into the cockpit, first standing on the seat pan of the right-hand seat before slipping my legs down into the space below the instrument panel. Once onto the seat, a member of the ground crew team follows me up the ladder and assists me with the business of attaching myself to the aircraft and the ejection seat, routeing and attaching the leg restraints (which go through the clip-on garters around one's lower legs) to the base of the seat, then attaching the PSP (personal survival pack – the dinghy and associated gear), PEC (personal equipment connector – which supplies air and oxygen and the pressurised air for the gsuit) and RT lead before routeing and securing the seat and parachute harnesses. Helmet and gloves on, oxygen mask attached, and by this time Mark is already busy climbing into the left-hand seat. By now it's clear that sitting in a Lightning is an intimate affair and underneath the endless straps, cables, rubber suit, helmet and oxygen mask, there's a distinct feeling of claustrophobia which is only increased when the huge and bulky canopy is lowered.

The biting cold of the Lincolnshire winds is replaced by warmth and a distinct odour familiar to any fast-jet pilot – a sickly but strangely reassuring mix of kerosene, oil, leather and vomit. With everything set to go, Mark signals to the ground crew and starts the first engine, resulting in the familiar hissing spurt of noise from the Avpin unit, and a slight tremor is felt through the seat as the mighty Avon rumbles into life at the first attempt. The instruments start to wind up and once Mark is content that the engine is safely stabilised, the second engine is successfully started, accompanied by another loud hiss and a growing rumble which is mostly felt rather than heard, although even with the protection of a very thick 'bone dome' there's no doubt that the engines are turning. Canopy down and locked, the wheel chocks are pulled way and at last we're free to go flying.

With a gentle hint of power, the brakes are released and we roll forward, performing a jolting curtsey as Mark checks the brakes for effectiveness. Then we turn right and begin to wind our way along the flight line past the assembled line of sister aircraft and out onto the long taxiway towards the northern end of Binbrook's main runway. The ride is comfortable but quite bumpy thanks to the Lightning's narrow wheels which are pressurised to an astonishing 360psi. Hardly surprising that I can feel every bump and crack in the concrete surface. Once the initial feeling of confinement begins to wear off, the view outside becomes more interesting and, although the Lightning's canopy framing is quite heavy, it's easy to see out to my right-hand side as far back as the aircraft's wing leading edge. The view to the left isn't quite so good, and apart from the very close shoulder-to-shoulder proximity of Mark my pilot, there's a lot of instrumentation and flying controls between me and the other side of the canopy frame. The view ahead however is very good, especially when my last flight had been in a tandem-seat Hawk, sat a few feet directly behind the pilot and a blast screen. By comparison, it feels as if I'm sitting on the very nose of the aircraft. Out to my right, I can see the famous crash gate where I'd spent so many hours with my camera, watching the Lightnings trundle by on their way to the runway The crash gate became what was probably the most popular and well-known aircraft enthusiast meeting point in the whole country, where people from all walks of life (some even coming from overseas) would gather simply to stand and look, fascinated and captivated. It's strange to see the crash gate from the opposite side, and it isn't particularly surprising that even on this grey, windy and cold afternoon there's still a visible crowd of onlookers, some giving us a wave as we go by. I wave back, of course. Just seconds later and we come to a halt at the holding point just short of the runway while Mark completes his run-through of the last pre-take-off checks and removes the safety pin from the seat pan between his legs while I remove mine and safely store it. The ejection seats are now live and once we have sufficient forward speed, we can use them if we need to at any stage, should things go very wrong, very quickly. Helmet visor down, seat straps good and tight, safety pin out, canopy locked, legs as far away from the control column as possible (the last thing I'd want to do is restrict its movement) and I confirm to Mark that I'm good to go as he talks to Binbrook's ATC to seek permission for line-up and take-off. We roll forward and turn left, out over the runway 'piano key' threshold markings and onto the centreline with a blaze of twinkling runway lights to our sides, stretching to the horizon. That's it – we're here and we're going flying!

Mark pushes the throttles slowly forward and there's a roar of muffled jet noise from somewhere behind us, combined with a growing tremor which rattles through the aircraft. The instruments show that the thrust from the two Avons is coming up to 80 per cent of their maximum power and the aircraft starts to nod nose-down slightly, the power pushing against the wheel brakes. But we're not hanging around – off come the wheel brakes and the nose perks back up as Mark brings the engines up to full power and we lurch forward as the roar of the Avon's thrust thunders against the airframe and makes a valiant effort to push its way through the canopy and my 'bone dome'. We're rolling forward and with another manipulation of the throttles from Mark, the reheat comes in and we suddenly push forward still faster with a surprising but smooth kick, almost like changing gear in a sports car. We're already racing down the runway, weaving very slightly either side of the centreline markings, and the raised barrier at the far end of the runway is already getting alarmingly near, and we seem to be bounding towards it at breathtaking speed – which of course we are

by now – and a quick glance at the airspeed indicator tells me that we're already way past 100mph. By the time I look up again, Mark is gently lifting the nose wheel off the runway, and almost imperceptibly the aircraft swiftly follows, rising smoothly and effortlessly into the air, and not a moment too soon, as the runway threshold flashes by just 50ft or so below us. The undercarriage immediately slams up (before the limiting speed of 250 knots is reached) and a couple of muffled bumps are confirmed by the 'three greens' popping off on the instrument panel. The Lincolnshire landscape is a few hundred feet below now and it's racing by with such speed that it's difficult to keep an eye on any particular detail close by as it would be gone as soon as you look at it. Far better to look straight ahead through the windscreen into the gathering gloom. Down goes the port wing and up comes the starboard wing as Mark rolls us to the left and hauls the huge brute round towards the east, before pulling us up into a gentle climb which pops us straight into the surrounding cloud. Suddenly the feeling of speed is replaced by a weird motionless sensation, suspended in a sea of greyness. The roar of the engines has subsided now (or more precisely we're pushing ahead of it) and inside XS458 there's a more comfortable, cosy feel of a gentle tremble from beneath our seats, combined with a steady rush of airflow hiss, competing with the buzz and murmur of the radio chatter. There's a lot to occupy my attention inside the cockpit, with a bewildering array of instruments and switches right in front of me, some of which I recognise, like the artificial horizon and compass, and of course the strange strip speed indicator that was fitted to later Lightnings. Even stranger is the throttle box, which instead of being close to my left hand, is over on my right side against the cockpit wall. I'm no pilot, but the idea of holding the control column with my left hand just seems bizarre, the more I think about it.

But enough of the inside – we're through the gloomy cloud layer now and out into the sunny skies over the North Sea, the Humber Estuary off to our left somewhere under the carpet of lumpy white-topped cloud below us. We're heading north and slowly creeping up to 18,000ft for a rendezvous with a refuelling tanker somewhere off Newcastle. The Lightning T5's lack of fuel capacity would normally mean that by now we'd already have to be thinking

ABOVE: Captured on camera just days before their retirement, a Lightning T Mk 5 twin-seat trainer leads a Lightning F Mk 6 single-seater in close formation over the North Sea. As can be seen, varying camouflage paint schemes were applied to the Lightning during its last years of RAF service. (*Tim McLelland*)

about heading back to Binbrook soon, but with tanker aircraft busy refuelling Tornado F2s ahead of us, there's an opportunity to fill up without going home. Mark introduces me to the radar display – a small CRT screen buried inside an almost comical rubber 'boot' which enables the viewer to stick his face against it and block out any distracting glare. I peer inside and it's certainly easy to see the screen display, but I'm not exactly sure what I'm looking at. Mark tells me that there's a 'blip' on the screen which is our tanker, a few miles ahead of us, but all I can see is patches of radar colour, blobs and specks, and nothing that looks much like an aeroplane.

Of course, I've watched too many movies and I'm almost expecting a beautifully clear screen with one aeroplane-shaped dot in the middle, but the Lightning's radar is a creature of the 1960s (even older in fact) and recognising the radar picture is undoubtedly an art which only the experienced eye can comprehend. I marvel at the very idea of making any sense of the radar picture and mention to Mark that it must be incredibly difficult to peer at his radar screen, fly the aircraft and intercept a target at the same time. He laughs and simply says that it's easy once you've learned how to do it, but I'm not so sure. He's telling me that we're closing in on the target now and that we should make visual contact very soon, so I take that as a good excuse to give up trying to make sense of the radar display and I peer out into the bright blue sky to find our target, and sure enough there it is, far in the distance, a tiny grey blur surrounded by even smaller grey speckles. We close in and the shape turns into a Victor from Marham, accompanied by a couple of Tornado F2s from Coningsby and two Lightnings from Binbrook which are busy topping up after flying some PIs (practice intercepts) further out to sea. Mark smoothly brings us alongside the port side of the formation so I can take a look at the refuelling procedure as the two Lightnings pull back and begin to descend while the F2s gently edge back and plug in to the Victor's trailing hoses, one on each wing. We don't stop too long though, and we begin accelerating ahead of the assembled gaggle in search of another tanker a few more miles ahead. Once again, Mark quickly brings us into a stern approach (and I'm still clueless as to what the radar screen is telling me) but just a few minutes later we're closing in behind a VC10 from Brize Norton, all alone and more than happy to fill our fuel tanks. After positioning alongside the VC10, we get clearance to move back to take our fuel. It all looks remarkably smooth and easy, just gently tucking in behind the VC10's port wing, slowly edging forward until the tanker's trailing hose basket is literally just a few feet away, twitching and wriggling in the slightly turbulent airflow in which we've managed to find ourselves. Mark eases us forward and the basket disappears to my left, behind the thick canopy framing, then reappears over Mark's shoulder, out to my left, and I lean forward as best I can against the tight seat harness straps, to see what's going on. There's a recognised technique for plugging into a basket which requires the receiver pilot to look straight ahead at the alignment markings under the tanker, moving forward until the probe makes contact, but Mark's an old hand and he's done this many, many times before. He's looking right at the basket and with some delicate tweaks of the control column he nudges the Lightning up and down, left and right, until he puts the Lightning's probe right in the middle of the basket and nudges forward. The probe makes contact with a discernable clunk and we're in, and the fuel starts to flow. It's only then that I look forward and realise just how close we've got to the VC10, the huge swept tail surfaces towering above

us. There's a sense of stillness with both aircraft flying in perfect formation and it's difficult to remember that we're all moving along at quite a pace, although I can see from Mark's concentration that things aren't quite so calm for him while he's keeping a careful eye on the position of the basket into which we're plugged, and the position of the VC10 just a few feet ahead and above. A few minutes later and we're full, and with a slight touch of deceleration we pull back and the probe pops out from the basket with another clunk, a spray of fuel vapour which briefly spatters across the windscreen, and a wiggle of freedom from the long, bolt-on probe. The VC10 majestically pulls away ahead of us and we slow and turn right before entering a gentle climb. We're heading east, out into the North Sea, and Mark applies power which is barely noticeable until I look at the airspeed indicator and see that the readout is moving. Mark tells me to ignore the airspeed indicator and concentrate on the Mach meter readout, which is more relevant at altitude. It's already moving past 0.8 and up to 0.9. Mark tells me to keep an eye on it as he selects reheat and as I feel a gentle push of thrust from behind us, the needle moves further along and seems to slow temporarily before creeping forward again across the magic figure – Mach 1.0. We're supersonic.

ABOVE: Delivered to the RAF in 1965, XR752 flew with No 23 Squadron at Leuchars, No 111 Squadron at Wattisham (as illustrated) and both Nos 5 and 11 Squadrons at Binbrook. After retirement it ended its days at Binbrook as an airfield decoy, before being scrapped when the station was closed. (*Tim McLelland collection*)

ABOVE: Lightning F Mk 3 XP751 joined the RAF in 1964, operating with Nos 23, 74 and 111 Squadrons. It then went to Binbrook and flew with the Lightning Training Flight and No 11 Squadron until October 1986, when it suffered an engine fire and never flew again. It became an airfield decoy and was eventually scrapped. (*Tim McLelland collection*)

You'd imagine that there would be some sort of noise or a bump or something fairly dramatic to tell you that you've broken the sound barrier, but inside the Lightning there's nothing remarkable at all to see or hear. Things look and sound just the same and the Mach meter just keeps creeping forward ever higher until Mark cancels the reheat and reins in the Lightning's phenomenal power. I'm imagining that somewhere down below, a lonely fishing trawler will have been treated to an unexpected sonic boom, but up here in XS458 it's just another routine part of regular Lightning flying. No magic, no effort. We slow down and turn back towards land and Mark takes the opportunity to show me a little of the Lightning's manoeuvrability, rolling and turning the aircraft, demonstrating that for a very old lady, she's not only fast on her heels but surprisingly nimble too. Then, much to my surprise, Mark asks me if I'd like to take the controls and before I've even stopped to think of what an inexperienced non-pilot civilian is supposed to do with a Mach Two interceptor, he tells me that I have

control and I sit there, slightly surprised and unsure what to do next. My first tentative steps to move the control column result in a few banks left and right and after just a few seconds it's easy to appreciate just how responsive the Lightning's controls are. The Hawk that I'd flown for a few minutes a few weeks previously seemed to handle very similarly, and yet the Hawk is half the size (and half the age) of the Lightning, but if I'd been flying blindfolded, I doubt if I would have been able to tell the difference. Mark promises me that I won't break anything (including ourselves), so I follow his instructions, pushing the control column over hard left, which immediately throws us into a rapid and disorientating roll, something which seems alarming at first but instantly becomes addictive, so back we go rolling the other way … and back again. But we can't play for ever. Having taken the Lightning into supersonic flight, the all-important fuel has started to dwindle again, and so we resume radio contact with our VC10 tanker and head back to the long, stately racetrack that the tanker has been holding whilst waiting for customers. Finding the lumbering tanker even with the Lightning's ageing radar equipment isn't difficult for Mark, especially when he also has the luxury of radar/radio vectoring to literally point him at our target, and it takes us just a few minutes to race back to the VC10 and slow ourselves down to come alongside and then take on some more fuel. Once full, we say our goodbyes (I give a friendly wave to the faces I can just see behind the tanker's cabin windows) and we turn back eastwards, descending towards an area of growing grey seascape which has emerged amongst the carpet of white cloud.

Down we go, through 15,000ft, 10,000ft, 5,000ft, and I have to make plenty of effort to keep unblocking my ears as the grey sea slowly resolves itself into greens, browns and blues, capped with white fringes of wave crests. We're still descending and the altimeter is showing us down to just 1,000ft and still slowly going down, until we smoothly level out over the North Sea at 500ft. Suddenly a small ship of some description sweeps by below us and then with nothing else in sight apart from a greeny-grey horizon, Mark takes us down still lower, right down to 250ft, and then down even more until the sea looks like it's going to come up and wrap itself around the cockpit. The throttles are going forward too as we hug the wave tops and in comes the reheat again, pushing me firmly back into the ejection seat. The grey wave tops become a blur and we're thundering forward, still accelerating, bumping and banging in the ridiculously rough air down at sea level, salt spray washing up the windscreen and the whole instrument panel shaking and rattling. Mark calls 'Pulling up, standby, three, two, one' and suddenly the world seems to turn upside down; I'm squeezed hard down into my seat – so hard that I can hardly keep my head up. It's almost as if someone has dropped an elephant on my head. I can just see out to my right a huge cloud of water vapour billowing around us and XS458's wings are actually moving – they're flapping and flexing and we're going up – straight up, vertically. The gforce stops and as the pressure in my gsuit slowly releases its grip on my body I can turn more comfortably to see the horizon at a crazy 90 degrees and the seascape rapidly receding far behind us. Another less brutal application of gforce comes back and the horizon slowly appears ahead of me through the windscreen now, but it's upside down, and with a neat roll the world turns the right way up again and we're straight and level, just a few seconds later, at 25,000ft. Quite a ride.

The show's over now, and we begin a long, gentle descent back towards Binbrook, back into the waiting cloud cover and then out underneath into the gloom over Lincolnshire, until the welcoming sparkling lights of Binbrook's runway appear ahead of us. We race back at 480 knots, thundering in across the airfield before making another very sprightly turn to port, with another fierce application of gforce which finally stops for good as we level out in the airfield circuit, flying northwards on the downwind leg, airbrakes out and the landing gear rumbling out into the air stream. Everything suddenly seems slow and rather pedestrian now, even though the instruments say that we're still motoring along at over 170 knots, and as I spot Grimsby's dock tower through the windscreen we turn left through 180 degrees round onto final approach and settle onto a gentle power-on descent, framed through the windscreen by a blaze of runway approach lights. The instruments read 160 knots now, but the sensation of speed starts to reappear as the ground gets closer. Everything looks fine until the very last minute as we streak over the narrow road that crosses the approach path – a place where I'd often stood awe-struck as the Lightnings roared overhead. The seemingly distant runway suddenly rushes up to us at breakneck speed and with a jarring thump Mark slams XS458 back onto the runway, cracking open the brake parachute which quickly starts to slow us down even though I can't see much evidence of the fact, or feel anything to suggest that we're ever going to slow down. But as the familiar crash barrier starts to look as if it's getting a little too close again, our speed finally settles down to something a little less ballistic and as the Q-Shed comes into view on the left-hand side we're back down to a gentlemanly taxi speed. Off the runway past the Q-Shed, brake chute dumped, safety pins back into the seat and back to the flight line where the canopy can finally be popped open and a welcome rush of cold Lincolnshire air allowed into the cockpit, almost as if to bring me back to reality. We're back, it's over and XS458 is comfortably at rest again, popping and creaking as the engines stop and things start to cool down. Mark climbs out onto the ladder and shouts, 'Was that okay?' I think it was more than okay.

It seemed only logical that having sampled the fascinating delights of the Lightning's capabilities, I should investigate the aircraft that succeeded it in RAF service. The Phantom was another thoroughbred that enjoyed a reputation almost equivalent to the Lightning, even though the F-4 was from a slightly later era. When I was first afforded an opportunity to fly in a Phantom I could hardly believe my luck, and finally I was standing at the foot of a Phantom cockpit's access ladder. From this angle, the Phantom looks barely capable of moving at all, let alone at supersonic speeds, or executing rolls or loops. It's a Chieftain tank with wings, a metal monster that has to be conquered like a mountain, assailed by ladder until the Phantom's very heart can be entered. Climbing to the cockpit isn't easy, not least because the cockpit is a very long way from the ground. At the top of the ladder, the experienced pilot could quickly climb into his cockpit, but for a novice passenger the route is slightly different. There are of course two cockpits, and one is positioned directly behind the other, each under its own metal-framed Perspex hood. The second cockpit was normally occupied by a navigator, although the RAF officer who worked in here did much more than simply navigate. He did indeed devote some of his time and energy to map reading and the operation of navigational equipment, but his main task was to operate the

LEFT: An unusual photograph illustrating three fighter types operated by the RAF's No 29 Squadron at the time of the unit's transition from the Phantom onto the Tornado F Mk 3. British Aerospace's Lightning test airframe was repainted in the unit's colours for this photograph. (*BAE Systems*)

Phantom's weapons systems. He spent most of his time peering into a radar screen, carefully making sense of a seemingly meaningless mash of signals, while selecting missiles and allocating them to their targets. The navigator directed the pilot onto the Phantom's targets and maintained a lookout for other marauding fighters. He monitored fuel, electrics, hydraulics and more. He was certainly much more than a mere map reader and even though pilots often referred to him in jest as the 'Man in Back' he was in fact a vital part of the Phantom's two-man team. By stark contrast, I was to be a mere passenger. Even getting to the Phantom's rear cockpit was a physical challenge, but from the perilously lofty ledge above the aircraft's gaping air intake, there is an opportunity to peer into the dark, shadowy cockpit and learn a little of what goes on inside there. Even a passenger cannot fly in the Phantom without being familiar with the interior of the rear cockpit, as there are countless switches, lights, knobs and handles that perform a huge variety of functions, some of which might cause a minor inconvenience if operated at the wrong time, but others too that have the potential to kill if they are not treated with respect. It's important to listen very carefully as the function of every switch is explained, even though it's far too tempting to simply 'switch off' and divert one's thoughts to the wider picture. As you kneel above the Phantom's rear cockpit, the massive bulk of the aircraft stretches out before you, and it's impossible

to avoid the temptation to just marvel at the picture. The long fuselage spine, the two monstrous intake trunks, the flat wings, turned up at the tips, the two down-turned stabilators, the towering tail fin … It's literally too much to take in.

My pilot directs me towards the aircraft's rear cockpit, where another of 92 Squadron's pilots is waiting to greet me. He then goes off to conduct a careful inspection of the aircraft's exterior, checking that all of the various safety covers and locks have been removed, and that there are no signs of any damage, fuel leaks, loose panels, or any other potential pitfalls. Of course, the ground crew have already ensured that everything is ready for flight, but history demonstrates that nobody is perfect, and a second look at everything is always a good idea, even if only for the pilot's peace of mind. Meanwhile, I clamber clumsily towards the rear cockpit and make my way inside, climbing down off the intake trunk and onto the ejection seat base, before stepping down still further onto the cockpit floor and settling myself into the seat. The ejection seat is already quite familiar to me as I've spent some time in another identical seat, learning how to route the various straps and fasteners correctly, and how to initiate the ejection sequence, should it be required. Having sat down, the first task is to reach forward and direct the leg restraint cords through the belts attached to my calves, and then plug each cord into its socket. The cords are loose but if the seat were to be fired, they would immediately pull tight and draw my legs into the seat base, hopefully ensuring that I didn't leave them behind in the cockpit as the seat rocketed outwards. The seat and parachute straps then have to be routed and clicked into their sockets, before being pulled tight so that I'm firmly attached to the seat back and base, virtually unable to move without first flipping the small handle that allows the harness to slacken a little for movement, when required. Although this makes things pretty uncomfortable, it has to be like this as the ejection seat is not some silly gimmick from a James Bond movie. It's a very complicated and potentially lethal piece of equipment that has the power to literally punch its occupant out of the aircraft in milliseconds, and rocket him away at phenomenal speed. If you're not firmly attached to the seat when it fires, you can be sure that you'll be injured, or possibly even killed. The all-important PSP lanyard is then plugged in, so that if I vacate the cockpit I'll take my pack with me, containing an inflatable dinghy and other 'goodies' that might help to keep me alive if I parachute to safety, only to find myself in the sea – or somewhere on land that's equally as ghastly. The lanyard also connects to a radio beacon that alerts the Search and Rescue crews, and although the system is obviously designed for an emergency, it's also important to remember that the beacon will also function no matter when the lanyard is pulled. It's not unknown for a pilot or navigator to climb out of the cockpit after a sortie, only to find that he'd inadvertently failed to unclip the lanyard, and a helicopter rescue crew was now on its way. The PEC is then plugged into its recess socket on the ejection seat, thereby linking me to the aircraft's oxygen and electrical system. The next task is to squeeze the bulky flying helmet onto my head and plug the RT and oxygen connections into the PEC. With these actions done, my 'helper' gives a friendly thumbs-up signal and disappears down over the cockpit sill. He's already removed some safety pins and set up some of the switches, but now I'm left alone and everything else within my rather restricted reach will now be my responsibility. By now I can hear signs of life from the front cockpit as my pilot has strapped in and connected his radio. With electrical power flowing into the Phantom, I can now hear the sound of muffled crackles through my helmet speakers, combined with breathing, comments to and from the ground crew and Air Traffic Control, and all of this is enough to satisfy me that I'm certainly not alone, even though the rear cockpit is a gloomy and solitary place. The interior of the shelter is already quite noisy, with roaring electrical generators pushing power into the aircraft, but now it's time to start the Rolls-Royce Spey engines and things are about to get far noisier – or at least I imagine so. In fact, with my helmet firmly fixed on my head and the internal head phones sealed over my ears, the engines start sequence isn't as exciting as I'd imagined, and the engine start is more of a feeling than a sound. Slowly, the aircraft develops a gentle rumble of disturbance, and there's soon no doubt that the Speys are turning,

especially if I push my finger between my helmet and ear to allow the true noise of the growling engines to reach me. With final checks complete, it's time to emerge into the daylight, and with a gentle nudge of engine power the huge machine edges forward and we roll out into the gloomy overcast, a welcome breeze drifting into my cockpit where things are already warm and stuffy. Out in the daylight I had imagined that the feeling of confinement would be less evident, but even without the shelter's all-embracing shadow, the Phantom's rear cockpit is still a dark and claustrophobic cell, with high-sided walls and a huge front panel of instruments, embellished with countless 'black boxes', pipes, cables and other assorted items that serve to block any decent view of the outside world. The front cockpit might as well be behind me, as there's no way of seeing it, hidden away ahead of the rear cockpit's clutter. Little wonder the navigators often referred to it as the 'coal face.' Chatter on the radio assures me that everything is good to go, and after performing a series of function tests that can be observed and confirmed by the ground crew (such as the operation of the flying controls), the Perspex canopies are closed and locked, shutting out the vestiges of engine sound and making the world inside the rear cockpit even more confined. We rumble forward away from the shelter and turn out onto the taxiway, the ground crew giving a final thumbs-up before they retire into the fuel-tainted warmth of the shelter. Inside the Phantom the ride along the taxiway is a quiet and comfortable one, the aircraft's landing gear riding the occasional bumps with the grace of a limousine. The last of the safety pins are now removed so that the ejection seat is live and capable of delivering me to safety, should anything go wrong. But the seat has to be treated with the greatest of respect, as it has the ability to both save lives and kill. The black and yellow grab handles above and below the seat will initiate the ejection sequence, throwing the seat and its occupant away from the aircraft at phenomenal speed before releasing its grateful payload to a gentle descent by parachute. But the initial ejection is not some James Bond-esque lurch into the air from a large spring. It's a brutal, rocket-powered punch that can badly injure the seat's occupant if he's not fully prepared and firmly positioned against the seat structure. It requires correct posture and a very hard pull of the handle to ensure a safe ejection sequence, and even though a gentle tug on the firing handle is unlikely to achieve much, it is always vital to avoid touching the handles … just in case. Sitting on a live pack of rocket motors can sometimes be a slightly unnerving experience, but at the same time it's reassuring to know that a reliable means of escaping from a lethal situation is just a tug away. With no other air traffic to delay us, we roll straight past the holding point where aircraft would normally await clearance for take-off, and we turn onto the 'piano key' threshold markings at the end of Wildenrath's light-studded and rain-washed runway, followed by our 'playmates' in the second aircraft that slowly tucks into position just behind us, ready to get airborne a few seconds after us.

With a final check that the safety pins are all removed and that the harness straps are all tight and locked, helmet, visor and oxygen mask all properly secured, I acknowledge to the muffled voice from the front cockpit that I'm ready to go. If anything goes wrong, the take-off will be aborted courtesy of the brake parachute or the arrestor cable hook. But if we reach a speed of 115 knots, we'll keep going unless something catastrophic occurs, such as a fire or an engine failure. At that sort of speed it will probably be safer to get airborne and try to solve the problem or eject, rather than try to bring a speeding Phantom to a safe halt on the ground. It is time to go. The two Spey engines now begin to demonstrate that so far they have merely been resting, and as the rpm gauges rapidly wind up to 80 per cent, the Phantom begins to shake and shudder, the instrument's visibility trembling as the power of the two Spey engines rattles through the airframe. Off come the brakes, and with a sudden lurch we push forward while the engines give a combined roar of enthusiasm as the throttles go forward to 100 per cent dry power. Then the afterburners blast into life, and despite taking more than ten tonnes of fuel and missiles with us, the mighty Phantom thrusts forward almost effortlessly and powers along the

RIGHT: Phantom FGR Mk 2 XV435 proudly carries the markings of No 92 Squadron, based at RAF Wildenrath in Germany. Towards the end of the unit's association with the Phantom, their aircraft were repainted with red tails in order to improve flight safety in the busy and gloomy skies over Germany. (*Tim McLelland collection*)

sodden concrete. I notice the increasing rapidity of the gentle bumps as the landing gear negotiates runway expansion joints, and the forward speed has already flashed upwards to a very impressive 100mph. Even though there is no neck-breaking Hollywood-style acceleration, there's also no doubt that we're getting faster and faster, so that in just a few seconds we're approaching an astonishing 200mph. The aircraft's nose slowly comes up (the stabilators are held in a fixed position by the pilot and the aircraft begins to rotate through 10 degrees when sufficient airspeed is reached) and almost imperceptibly, the huge warplane lifts into the air and powers into the gloom, the two spikes of afterburner flame punching a hole in the low-hanging cloud behind us. Inside the Phantom there's a satisfying but unremarkable mix of airflow and engine noise, but there's nothing to really emphasise just how fast we're moving or how much noise we're making. It takes a leap of consciousness to imagine the ear-splitting roar that is rolling across the fields below as our grey fighter plane makes its way into the clouds. The landing gear, flaps and slats growl and clunk into their retracted positions and the reheat is cancelled, while we continue our acceleration into the sky. The cloud and rain have already enveloped the aircraft before we even pass Wildenrath's airfield perimeter, and by now there's nothing but an outside world of grey to contrast with the even darker cockpit interior, where only a scattering of glowing switch captions break the monotony of black panels and black instruments. But in no time at all the greyness begins to lighten and the gloom slowly lifts until, with a sudden burst of light, the Phantom pops through the top of the cloud cover and soars still further into a beautiful, crisp, sunny sky. The visual sensation is a joy, but just feet away in the front cockpit my pilot is far too busy to admire the view. He is busy talking to the local air traffic people who are doing their best to vector the aircraft away from Wildenrath and off towards our intended training area. Climbing through 20,000ft, I can hear that our wingman is making his way along the same path towards us, and I fumble for the seat height adjustor to try and get a better view. The seat gently whirrs its way upwards until I feel my helmet make contact with the cockpit canopy, the rumbling vibrations from a combination of engine revolutions and airflow buzzing through my head. I lower the seat a fraction to avoid direct contact with the canopy and reach out to

my left to see as much as I can, out over the port wing. Sure enough, there's the second Phantom, seemingly motionless beside us, glinting in the winter sunlight. The sight of the magnificent Phantom is a thrill, high above the clouds in the environment for which it was designed. But this is not intended to be a sightseeing joyride for me. It's a training mission for the RAF crews, and after an all-too-short transit in the sun, it's time to descend into the gloom again, to begin the business of flying the Phantom as a fighting machine.

Back in the all-embracing gloom of the cloud cover, I'm disappointed to learn that our aircraft's radar is steadfastly refusing to function, and for the rest of this sortie we'll be obliged to fly what is in effect only a representation of a more normal mission. In reality it's already unfolding as a very normal sortie, as the Phantom's radar was often found to be unserviceable, especially as each aircraft got older and its equipment began to show its age. But the RAF's crews are resourceful, and losing radar has never meant that a sortie had to be called off. There's still plenty that can be gleaned from the flight, not least the all-important opportunity to practise the interception of targets through direction from both a ground controller and our wingman. Sitting in the Phantom's rear seat, I would have been busy keeping a track of the aircraft's progress by map had I been a navigator, but as a civilian, even a carefully marked out map is of little use without constant updates on our position from my pilot, who (thankfully) has the skills to fly the aircraft and navigate us. Had I been a navigator, I would also have been talking to the ground controllers, taking into account the needs of the mission whilst also accommodating the many pressures of peacetime training, which require minimum heights to be observed, restricted areas to be avoided, sensitive areas to be dodged when possible, and confliction with other aircraft to be accounted for. During the darker years of the Cold War, West Germany's vast, sprawling landscape gave the impression of spaciousness, but for the military aviator it was in fact a busy and potentially dangerous area, and always occupied by many fast and low-flying NATO warplanes operating from the countless airfields scattered across this region of Western Europe. But from my position buried in the cockpit, there's little to see outside other than a wall of cloud, which slowly thins to reveal a miserable patchwork of green, brown and grey fields, punctuated by scatterings of brown-roofed houses that seem to be shivering and cowering under the overcast. It certainly isn't the panoramic, sun-kissed vista that might have been hoped for, but it's at least another very realistic illustration of just what a typical training sortie was like. The weather was often foul and the Phantom crews were entirely accustomed to battling through rain, snow, thick cloud, thunderstorms, in fact just about anything that Mother Nature could throw their way.

After making a few adjustments to the radio switches, I turn my attention outside again and notice that we've now got closer to an industrial area, the sodden fields having mostly been replaced by grey buildings, roads, high-tension cables and railway track. A voice from the front cockpit confirms that we've brushed past both the Netherlands and Belgium and are now skirting the industrial perimeter of Aachen, heading south, with the sprawling city of Cologne off to our left. A few minutes later we're entering our designated CAP – a Combat Air Patrol – out over a stretch of countryside near Koblenz. Flying a CAP is of course entertaining for a passenger, but for the Phantom's regular crew it's a tedious activity, occasionally interrupted by sudden periods of frenzied action. We divide our two-ship formation into individual elements, with each Phantom flying a long, straight leg of a huge racetrack pattern. One leg of the racetrack reaches out towards Koblenz, aimed directly at the distant border with East Germany and the Warsaw Pact forces that would (in a wartime scenario) be heading towards us. This is referred to as flying 'hot' while the other aircraft reverses along the racetrack (flying 'cold'). This arrangement ensures that at least one aircraft will always be directing its radar towards the anticipated threat direction, the nose-mounted radar dish sweeping 60 degrees left and right at a rate of around 120 degrees per second. Of course, it's a procedural exercise that can (at least if no 'trade' is to be found) become tiresome in the extreme, the pilot being required to maintain a carefully controlled route at low level, and often in depressingly poor weather. With no functional radar to play with, I can only occupy myself by looking outside, watching the fields and houses rush by just 500ft below us while we maintain our patrol. I'm surprised to notice that even though people can occasionally be seen below, we don't seem to be getting much attention as we roar by.

Maybe it's a symptom of speed, and perhaps a few more heads turn skywards a few seconds later when the thunder of our monstrous Spey engines finally makes an impact, but I can't help thinking that maybe the locals are simply accustomed to warplanes, and they have long since stopped bothering even to look at just another Phantom going about its business. Like other NATO forces in the region, the Phantoms were sometimes referred to as the 'sound of freedom' and it's certainly true that they did represent a nuisance that the Germans accepted and tolerated, as a necessity of the Cold War.

Thinking about such matters occupies my thoughts for a minute, but my attention is soon directed elsewhere. Our steady sweep across the countryside is about to end abruptly, as a call over the RT reveals that we're no longer alone. While we've been maintaining our patrol, a pair of marauding fighters have been looking for us and now they've found us. Suddenly, the world seemingly turns upside down, the engines rumble and as the afterburners kick in, the mighty Phantom surges forward, rolling to the right and into a tight turn. My anti-g trousers suddenly seem to explode, instantly filling with high-pressure air and clamping my legs and stomach with a ferocious intensity, but this is only a minor distraction from the gforce that's piling down on me as we haul round a tight turn, huge sheets of cloud billowing and rippling above the wing tops, and long streams of vapour trailing out from the waggling wingtips. In an instant the brutal gforce stops as we roll wings level and then haul left, and on comes the gforce again, more and more, then a slight relaxation as we start to head upwards, then down again, still turning, rolling, and fighting to shake off the simulated missiles that are being directed against us. Although my pilot is working as hard as anyone can, life in the back seat is no fun either, even for an idle passenger. There's nothing to do other than grit one's teeth and hold on as the fight progresses, whilst trying to make sense of the world as it rushes past to the left, to the right, over my head, all over the place, until it seems easier just to look straight ahead at the dark instrument panel, where at least the dials and switches seem to stay still, although I have a hard time keeping my head steady enough to tell what any of them are actually indicating. They tell me that for most of this brutal fight we're still down in the weeds at around 500ft or less, thundering along at speeds of 500mph or more. I can

LEFT: Phantom FG Mk 1 XV569 during a visit to Alconbury in 1980, whilst assigned to No 111 Squadron based at Leuchars. This aircraft first served with the FAA and was subsequently transferred to RAF service when the Royal Navy relinquished its Phantom fleet in the 1970s. (*Tim McLelland collection*)

do nothing but wait for it all to end, and marvel at the dedication of the navigators who are obliged to sit in this hell-hole and do a deadly serious job. They tell me that it's never easy, and some of the most experienced navigators admit that although they love their job, it can also be a pretty miserable task at times. When the Phantom is engaged in a fight, the navigator's cockpit is a notoriously unpleasant place to be. The view outside is limited even at the best of times, and the walls of black instrumentation offer nothing but data and a persistent smell of rubber and oil that seems to permeate the even fouler smelling rubber oxygen mask that is clamped to my face. Peeling the mask away for a welcome breath of fresh air is inevitably a disappointment, as the cockpit is anything but fresh. It reeks of kerosene and oil, mixed with rubber and sweat, plus an unmistakable hint of human vomit. The navigators make no secret of the fact that throwing up is just a regular part of life in the Phantom, and as I sit there, sweating and gasping, I begin to feel as if I might need one of the sick bags that I've carefully zipped into the legs of my anti-g trousers. It takes a lot of effort and a lot of deep breaths to draw my mind away from the growing urge

to puke, but at least the fight has been called off as both we and our interceptors are starting to run low on fuel, and we begin our transit back to Wildenrath.

Out off the left wingtip the unmistakable shape of a Lightning appears from the patches of low cloud and closes up in formation to follow us north, before heading off in search of a refuelling tanker. The sight of the RAF's iconic interceptor just a few feet away is enough to take my mind off any thoughts of being sick, and I watch the magnificent grey beast jiggle and jostle for position beside us until, with a final burst of condensation streams, the Lightning peels away and roars off into the cloud, leaving us to head home with our wingman trailing us a couple of miles behind. We race north towards Wildenrath through the driving rain and curtains of cloud, while I try to gather my composure after what has been a pretty exciting few minutes. My head is still spinning and even though we've now settled into a steady attitude, I'm hot, sweaty and surprisingly tired, considering I've never even left my seat. It's a welcome relief to hear the laconic tones of Wildenrath's air traffic controller crackling over the radio, providing us with weather information, airfield data and directions through the gloom until, just visible through the murk, a fuzzy blob of orange-white light appears and we turn gently to head towards it. Although it's difficult to see much ahead of me beyond the cluttered cockpit, an occasional glimpse into the distance assures me that the blob of light is swiftly growing into a stretch of runway lights, and after my pilot calls 'initials' (indicating to the air traffic controllers that we're rapidly approaching the airfield), the controller replies with a final update of the runway in use, the wind direction and speed, and a clearance to run in and break left into the airfield circuit. Seconds later we're over the airfield and at 500ft the scattered aircraft shelters are spread out below us like fingers reaching into the surrounding trees. With the airspeed indicator at 450 knots we roll left and pull hard into a 180-degree turn, the throttles closing and the airbrakes popping out, as the gforce suddenly piles on again through the turn. It's a relief when my anti-g trousers gently deflate for the last time as we settle on the downwind leg of the airfield circuit, our wingman racing in behind us. The airspeed indicator drops below 250 knots and a sudden clunk and shudder indicates that the landing gear and flaps are extending. Speed continues to drop to just above 150 knots and with the flaps fully down and the landing gear dangling, we gently roll left into a lazy turn onto final approach. We settle on approach half a mile from the runway at 300ft, and now my pilot plays with the throttle and control column to give us a steady 19.2 units of AOA – the ideal approach for the Phantom. By stretching left and right I can see the runway lights ahead of us, but it's a struggle to see much at all straight ahead. For a second I wonder how instructors ever managed to land the RAF's 'twin-stick' Phantoms from the rear cockpit, but I guess skill and practice can compensate for even the most unsuitable conditions. We race towards the runway and as the runway 'piano key' markings streak by we thump back onto the concrete with quite some force.

As a naval aircraft, there is no need to flare for touchdown and the correct procedure is literally to fly the Phantom at the runway, and the sturdy landing gear takes care of the rest. Once the wheels make contact the brake parachute is released and almost immediately it begins to have a noticeable effect on our forward speed. Once the airspeed indicator slips below 100 knots the wheel brakes are also slowly brought in, and after just a few seconds we're rolling gently along Wildenrath's runway at little more than taxi speed. Even though there's a steady drizzle drifting across the airfield, both my pilot and I can't wait to crack open the Perspex canopies to allow a delicious flood of cold, clear air to rush into the hot, stinky cockpits. The ultimate ride (at least for me) is over, and all that's left to do is to return to the shelter complex where the engines can be shut down, and after a post-flight inspection the mighty machine can be winched back into the dry and readied for its next flight. For the aircrew a detailed debriefing is to follow, but as a passenger I can take my time to gather my thoughts, relax, and thank everyone for allowing me to sample the mighty Phantom at first hand.

9

SHOWMANSHIP

The 'Blue Angels' aerobatic display team was formed in 1946 and since that time its performances have thrilled more than 250 million people. During their 1992 European tour, they confidently expected to add millions more to their ever-growing band of admirers. The European tour was aimed largely at the central and eastern areas, with the team scheduled to fly sixteen shows, but the only British appearance was at RAF Finningley's 'Battle of Britain' show in September. The first of sixty US shows in 1992 took place on 14 March at Naval Air Facility El Centro in California, where the team had been busy training since early January, having deployed there from their home base in Pensacola, Florida.

The 'Blue Angels' relocate to El Centro every year for pre-season winter training. Anyone who has spent a happy vacation in sunny Florida might wonder why the squadron has any need to leave its home base for almost three months, but the reason is largely one of weather predictability, as even Florida can suffer from periods of cloud and rain, which would obviously disrupt a hectic training schedule. Additionally, the move enables the pilots to devote their time entirely to flying, rather than having to deal with family and domestic problems that inevitably arise.

Situated deep in the sun-baked California desert, El Centro is certainly remote. Roughly three hours' drive from San Diego, the record-breaking 1992 winter storms did not reach this far, and the training season was entirely unaffected. El Centro is essentially a huge expanse of dust, with just a few mountains to disrupt the horizon and the long Mexican border only 5 miles away. Clearly it's a perfect place for an airfield, and an ideal location for training activities, which is why weapons firing and formation aerobatics are common sights in the area.

When I arrived at the Naval Air Facility, the familiar shape of an RAF Hercules on the flight line could be seen. Further investigation revealed that the RAF Falcons parachute team had also learned to appreciate the advantages of desert deployment and, like the 'Blue Angels', they too now make winter visits to the base. The 'Blue Angels' were already preparing for their first flight of the day as I entered the crew's facilities, even though it was only 7 a.m. Two display rehearsals are normally flown daily at El Centro, from Monday through to Saturday, the first one at around 7.30–8 a.m., and the second at around 11 a.m. During the early weeks of winter training, the main four-ship formation will fly as a separate display unit, with two solo performers making their appearance after the main group each day. Towards the end of the deployment, the two elements combine to form the basis of the full display profile. While most practice 'hops' take place over the nearby Shade Tree weapons range, the final pre-season rehearsals are flown directly over the airfield at El Centro.

During the first day of my visit, the Commanding Officer, Commander Greg Wooldridge, arranged for me to fly on a demonstration profile flight in one of two F/A18B aircraft that are normally assigned to the

ABOVE: Over the top of a wide, lazy loop, the lead aircraft is just a few feet ahead of the trailing aircraft (from which the photo is taken) with landing gear deployed. Maintaining tight formation in such exuberant manoeuvres is far from easy, even though the 'Blue Angels' make it look simple. (*US Navy*)

squadron. The aim of this flight would be twofold: firstly, it would enable me to see exactly the kind of demonstration that is given to selected media representatives and a handful of celebrities each year; secondly, it would enable the team to establish whether I was capable of withstanding the rigours of Hornet display flying, prior to joining the full six-ship formation.

Lt Dave Stewart was the 1992 'Blue Angels' narrator, and as pilot of aircraft No 7 (the twin-stick F/A18B), was also responsible for all media demonstration flying that year. He is a graduate of the famous US Navy Fighter Weapons School ('Top Gun'), and has over 3,000 flight hours to his credit, including 290 carrier landings. Having been introduced to Dave, I then met his crew chief Greg Braden, to be briefed on the safety aspects of flying in the rear seat of the F/A18B. Kitting out for a flight with the 'Blue Angels' is a relatively simple task when compared to other forms of fast-jet flying. Unlike many of their counterparts, the 'Blue Angels' pilots do not wear anti-g garments of any sort, despite the fact that their day-to-day flying is very much in the 'high performance' category. Consequently, my flying clothing was limited to just an all-blue one-piece overall, which for a demanding workout in the hot desert was particularly fortunate. The absence of gsuits is compensated for by a regime of regular exercise, in which all pilots participate. They lift weights and they run, almost every day, and at this stage of my visit I was beginning to wonder whether I'd regret being a fairly non-athletic type!

With the morning schedule of display practice complete, it is my turn to be given a taste of the 'Blue Angels' in action, and out on the flight line Greg eases me into the rear seat of F/A18B 161932, better known as 'Blue Angels' No 7. The strapping-in procedure is very straightforward for a passenger, as the crew chief takes care of the entire operation, including the fitting of the additional lap harness fitted to 'Blue' Hornets, indicative of the regular doses of inverted flying inflicted upon the occupants. Once comfortably seated on the British Martin-Baker Mk 10 ejection seat, I'm handed a blue flying helmet equipped with a boom mike, as the team has no need for oxygen masks, flying exclusively at low altitudes. Although their F/A18s are clearly not painted in grey camouflage, the 'Blue

Angels' Hornets are in every respect typical USN/USMC airframes. The only major modification made to the aircraft is the removal of the 20mm cannon and ammunition tanks, which have been replaced by a smoke generation system, feeding into the exhaust of the port engine. Otherwise, the 'Blue Angels' machines are essentially standard products. Having settled into my seat, Dave arrives to get the flight under way. Although it's customary to see the pilot performing a walk-around check of his aircraft, the 'Blue Angels' pilots delegate this task to the crew chief, having total trust in their professionalism. Once Dave is strapped in, Greg disappears from view down the port side of the fuselage, swinging the extended ladder back into its housing in the wing leading edge extension (LEX). Inside, the aircraft begins to come alive, as the internally mounted APU (Auxiliary Power Unit) feeds life

ABOVE: Naturally, a US Navy F/A-18 Hornet would never be seen in landing configuration whilst flying upside down in a loop, unless it was being flown by the 'Blue Angels'. Their beautifully maintained and finished aircraft show no indication of the fact that all their Hornets are much-used machines that have already flown countless punishing hours with front-line US Navy units. (US Navy)

into the electrical systems. The RT conversation between the occupants of front and rear cockpits is interrupted by the electronically generated female voice which recites a variety of warnings: 'ALTITUDE … ALTITUDE … FLIGHT CONTROL … LEFT ENGINE FIRE …'   While the pre-taxi checks continue, I check the height of my seat by operating a switch on the seat base. The crew chief has already advised me to ensure that there's at least a fist's width between the top of my helmet and the canopy, to ensure a comfortable ride when the action starts. In front of me, the high-tech Hornet displays are already beginning to look more than a little confusing to my untrained eye. The on-board computers are busy running through a BIT (Built-In Test) programme. With the engines turning, the three MFD (Multi Function Display) screens begin to illustrate the bewildering amount of information that can be made available to the pilot. On the right-hand screen, the BIT programme continues, listing a range of aircraft functions, together with the appropriate status caption. The weapons stations layout can be displayed on the MFD screen, although, of course, our display configuration is completely clean. Dave continues to run through his checklist on the left MFD, and although much of the information means little to me, I determine that the aircraft has a 7g limit, although this figure will rise to 7.5g as our 9,500lb of JP5 fuel is burned off. In practice the team pilots don't regard anything under 8.1g as an overstress, however. The canopy is closed and locked, the wings are confirmed as being unfolded (or 'spread' as Dave says, thanks to his Tomcat background). Trim is set at 12 degrees nose up, the flaps at half setting, hook is raised and locked, and we begin our taxi to the active runway, the undercarriage causing the Hornet to pitch and roll gently in a distinctly nautical fashion. El Centro's air traffic controller gives us the current weather conditions: 'Weather at two-two-one-one zulu, sky conditions better than five-thousand-five, the temperature eight-zero. Wind zero-two-zero at seven knots, altimeter three-zero-one-one, pressure altitude two-one-zero, depart runway zero-eight.' On the left MFD I now have a repeat of the front cockpit's HUD, which gives a reading of 48 knots airspeed (the lowest recordable figure) and a −30 figure for our altitude, indicating that El Centro is some 30ft below sea level. The vertical velocity figure changes randomly as we bounce along the taxiway, with a runway alert van following us, carrying a

crew of technicians who would be instantly on hand to rectify any last-minute problems. It's the squadron's proud boast that they have never been forced to cancel a display because of mechanical failure, and judging by the amount of time and care the maintenance crews devote to their aircraft, it's hardly surprising.

Out onto the runway we pause briefly, the tower having cleared us to make a maximum-performance take-off. The two General Electric F404 turbofans are run up to 85 per cent power and there's a muffled rumble of jet noise from behind my seat, the airframe shuddering slightly, in a gentle nose-down attitude, held against the brakes. Everything is looking good, and after a word from Dave, 'You all set?' I confirm that I'm happy to go and Dave releases the brakes, simultaneously selecting full military power on both engines. Just a couple of seconds later the afterburners kick in and give our forward motion a hefty push, and with 32,000lb of thrust now behind us, the Hornet is swiftly airborne at 125 knots, less than 2,000ft along the runway and just ten seconds from brake release. 'We're going flying,' confirms Dave, as the gear and flaps retract. Maintaining a 50ft height over the runway, the nose is held down until everything is tucked away safely, and at 350 knots we pass over the end of the runway threshold, some thirty seconds from brake release. With a 'STANDBY, STANDBY' from Dave, he makes a 5g pull, and a couple of gut-wrenching seconds later we're climbing vertically away from El Centro's runway, powering straight upwards into a deep blue sky. The take-off looks impressive from the outside, and I can confirm that from the inside it's even more remarkable, the crushing pull-up being instantly replaced by a rocket-like zoom straight into the stratosphere. Just fifty seconds into the flight we roll off the top of our climb, with the runway complex far below rapidly sliding away behind my shoulders. The climb is terminated as we head out towards the nearby Shade Tree range, and the landscape below appears to be largely featureless, apart from Superstition Mountain in the distance, which makes a great reference point, and the equally obvious Salton Sea out to the north. The range controller clears us into the area and although we're still within visual range of El Centro, a look down at the seemingly endless desert floor reveals a collection of huge circle markings, indicating the position of various bomb/gun targets that have been used here for many years. Hidden amongst these bizarre markings there's a long furrow, marking the position of the 'Blue Angels' display line, used for practice displays. Situated close by is a small white trailer which represents the display centre point, and some 1,500ft behind this, there's another trailer from where the maintenance officer (Lt Kevin Fischer) and flight surgeon (Lt Cdr Pat Spruce) carefully scrutinise each flight, making detailed notes of all the manoeuvres for post-flight debriefing. Each display run is also filmed on video for the same purpose, and as team narrator, Dave would also normally be there too, running through his show commentary.

Having taken a panoramic look at the rehearsal site, we gently descend for a closer look, making a low pass along the display line at just 100ft. Once complete, it's time to begin a sequence of display manoeuvres, and we pitch up through 25 degrees before entering a lazy barrel roll. It looks easy enough, but Dave reminds me that there would normally be another Hornet just 3ft away from each wingtip, and another almost directly above my head. Next comes a look at turning performance, and as we bank into a left-hand turn, the gforce begins to pile on, starting with just 2g ('As you can see, it's not much more than normal'), then increasing to 3g. At this stage, had I been wearing the usual gsuit, it would have inflated, but without it I'm left to perform my own anti-g operations. Greg, our crew chief, has given me plenty of advice on what to do if things get tough, the basic aim being to keep as much blood as possible from draining towards my feet. As the gforces increase, progressively smaller amounts of blood can be pumped to the brain and, unless the acceleration stops, I will eventually lose consciousness. The way to avoid this is to tighten the leg and stomach muscles as much as possible, in order to keep blood in the upper body. The result tends to be a great deal of squeezing and grunting, and indeed Greg's advice is to grunt the word 'hook' slowly (it comes out more like 'heeuuuch' however). This 'hook manoeuvre' was once graphically described to me as being rather like straining on the toilet, and it is indeed a strain – the rear seat of a Hornet is no place for politeness. Rolling right, we enter a sustained 4g turn and things get even tighter and sweatier.

The planned loop profile follows, and with a 4g pull we climb up into the deep blue Californian sky, easing off the gforces at the top, as the (inverted) horizon comes into view. Down the other side of the loop, and a 4g pullout at the bottom enables us to pick up traces of our own white smoke trail, which indicates 'a healthy wind blowing us back over the flight line'. From the loop we then enter straight into another climb for an Immelmann manoeuvre, rolling out from the half loop on the top, my left MFD displaying 10,900ft at the top. Things are starting to get hot inside the cockpit, but the action continues as Dave initiates a SplitS manoeuvre, rolling us over onto our backs, before pulling 4g into a vertical descent towards the desert floor. The MFD shows a 20,000ft-per-minute rate of descent, increasing to 30,000, before we haul upwards into level flight. I check the MFD again, and it confirms that although the nose is well above the horizon, we're actually still descending quite rapidly, the speed and inertia still forcing us downwards. Dave warns me that our Hornet is likely to 'shake and rattle' during this demonstration, and it lives up to my expectations, bucking and protesting as we gradually exchange the downward velocity for a more comfortable forward one again. Further shimmying and shaking is predicted for the next profile, this time a low-speed pass at a high AOA. The speed brake is deployed, and airspeed slowly bleeds away down to 125 knots, by which time the automatic trailing and leading edge flaps are well extended to keep our disturbingly slow Hornet airborne. We progress along the display line at this pedestrian speed, the nose pointing firmly upwards, leaving me with a peculiar sensation not unlike falling off a horse. Despite being firmly secured to the seat, the very unusual attitude and the remarkably slow speed give you the distinct impression that you're likely to slide off the back of the great beast at any moment. Accompanying this sensation is the pungent smell of re-ingested exhaust gas, reminiscent of another time when I'd hovered in a Harrier. The conditions don't last too long, however, as Dave then pushes the throttles forward into reheat, and without altering our attitude we're suddenly making a very powerful climb, straight ahead. Having looked at low speed, the opposite end of the performance range now follows, as we retrace our path back to the display line. Back at low level, Dave invites me to push the throttles forward to full military power, warning me that with a combination of wind and desert heat, we're 'likely to get bounced around'. We're reading 320 knots on the MFD, and with a further push on the throttles I select afterburners, pushing us firmly back into our seats.

'Okay, we have 500 knots now, you can see our shadow racing along the floor on our right … 600 knots … Mach point-nine-two … Mach point-nine-five … got to throttle back here …' and up we go into a steep climb, at just under the speed of sound.

With my forthcoming formation flight very much in mind, Dave next flies a vertical pitch recovery manoeuvre, which requires some pretty severe turning at 7g or more. I gasp and grunt my way through the pull, as my peripheral vision starts to fade, and I recognise the infamous 'tunnel vision' that leads to blackout and G-LOC (g-induced loss of consciousness). It's hard work for a passenger, but great when it stops! The MFD confirms that we're manoeuvring at 7.5g and, yes, the mighty Hornet's wings really do flex when that kind of pressure is applied. Dave enquires if I'm all right and once he's satisfied that I'm still conscious, I'm invited to try some manoeuvres for myself. After a couple of careful 4g turns to the left and right, I try a snap roll, and the temptation to utter a cheesy *Top Gun* (the movie, that is) style 'YEEHAA' is just too strong to resist. Another snap roll, and then a loop, with some careful tuition from Dave, guiding me through the 4g pitch-up, and my fairly acceptable wings-level climb over the top. As I pick up the horizon I can see our smoke trail, which makes the task of keeping the loop straight much easier. 'A real good job,' comments Dave from the front seat, and for a non-pilot it may well have been, but I'd hate to try the same manoeuvre with another Hornet just inches away from me. Straight and level again, I try flying the aircraft with the display-standard 15lb down-spring fed into the pitch control system. Although I've been warned that the aircraft will want to pitch down, I'm surprised to feel just how heavy the controls have suddenly become. It's hard work to fly the Hornet with this input (one is almost tempted to use both hands), and yet the 'Blue Angels' fly every display in this mode, in order to improve the responsiveness and crispness of the formation handling characteristics of the aircraft ('It takes out the sloppiness'). With Dave back in control we climb into a zero-g pitchout, giving us ten seconds of delightful 'spaceflight' as we balloon over the top of our ascent.

But the fun is now over, as the range controller asks us to vacate the area in order to allow two A6 Intruders to deliver their bombs. We head out toward the Salton Sea, awaiting permission from Yuma control to enter another box of controlled airspace to complete our flight. There's time to listen in on Yuma's air traffic, and the area is evidently very busy, with a variety of Intruders and Skyhawks using the weapons ranges, together with Tomcats and Eagles flying air combat manoeuvres. In fact, the airspace is too busy to allow us to complete the demonstration profile, and so we fly the final inverted manoeuvre at high level, instead of being at a more representative 1,000ft. With a 5-degree pitch-up, we roll sharply over onto our backs for a brief taste of inverted flying. We hang in our straps for ten seconds, roll out, and, after reversing the Hornet's direction, we flip over once more, snapping my head in the opposite direction. This time it's a long twenty-two seconds run during which there's plenty of time to become familiar with this rather unusual way of looking at the world. Once you learn to ignore the feeling that you're going to fall, it's actually quite enjoyable. The 'Blue Angels' F/A18s have been modified to allow a maximum of forty-five seconds of inverted flight, and as fuel pressure drops, there's a warning caption to give five seconds' notice of an impending flame-out. Time for one final manoeuvre, flying inverted-to-inverted, so over we go once more, the horizon flips around to our alternative position, and we hang, seemingly motionless, for ten seconds before making a 360-degree roll back into inverted for another ten seconds. 'There's just all sorts of fun to be had when you're flying,' comments Dave from the front seat. With this complete, it's time to head back to El Centro, and we thunder into the airfield circuit, making another airframe-buffeting 6g turn onto the downwind leg. The landing is also exciting, as Dave demonstrates a typical carrier approach. It looks easy, but as he explains, there's plenty to deal with at this critical stage in the flight. Apart from keeping on the runway centreline, it's important to maintain the correct approach speed, checked by a constant monitoring of the HUD angle of attack (AOA) bracket which appears on the display once the landing gear is extended. This has to be kept within the precise confines of the flight path marker, and the task is far from simple as it's not just a matter of moving the control column. Changing speed and attitude has to be a careful combination of stick and throttle. Additionally, Dave also has his eyes fixed firmly on the airfield optical landing system. Better known as the 'meatball', the device is basically two rows of green lights with a white ball of light appearing above (meaning you're too high) or below (too low). Even on land the task seems like a great deal to handle, and I could only imagine how much more difficult it would be to perform the same task on a pitching carrier in bad weather. The 8,000ft runway rushes up to meet us, and with a synthetic cry of 'ALTITUDE … ALTITUDE' from the computer, the Hornet thumps back onto the concrete. Had we been on a carrier, the eye-popping wire arrest would have followed, but we're safely back on land, gently rolling down the runway, before heading back to the flight line.

With the awesome experience of the Hornet orientation flight behind me, the next day I was invited to join Lt Cdr Pat Spruce, the 'Blue Angels' flight surgeon (better known as the Doc) on his daily trip out to the Shade Tree weapons range, to witness the morning's display rehearsals. After driving some considerable distance through miles of featureless desert, we arrived at the site of the two white trailers, which I had glimpsed many times from various altitudes and attitudes the day before. The team's maintenance officer (MO), Lt Kevin Fischer, was already busily setting up radio communication with the first four-ship formation, which was visible in the distance, just getting airborne from El Centro. A few minutes later, Dave Stewart arrived (his personal call sign being 'Hoops', which has been modified to 'The Hoopinator' by his colleagues and those who have had the pleasure of flying with him). The Doc communicated an update on the local wind speed and direction to Cdr Wooldridge, and the distant specks on the horizon rapidly grew into the familiar shapes of four Hornets, running in over the centre point in perfect formation, trailing white smoke. The following thirty minutes were occupied by a continued series of formation manoeuvres over the display line, every second of the performance being carefully monitored by the Doc and MO. Some comments were radioed directly to Cdr Wooldridge, while other points were written up for later discussion. Another member of the ground team stood forward of the trailer and cars, filming the four-ship's progress on video. Slightly further away from the action, Dave paced back and forth, concentrating hard on a huge pile of cue cards containing the entire commentary for both the full and bad weather display – the daily trip to the desert is a good place to mentally tie the narrative to the aerial manoeuvres before giving his first presentations

to the listening public. Just a few minutes into the display, the four-ship formation became a three-ship, as Lt Pat Rainey (No 4 slot pilot) took his aircraft back to El Centro. Today he was flying the twin-seat No 7, with a naval officer VIP riding in the back seat. Despite the VIP's shakedown flight and considerable naval experience, he had evidently uttered the key words 'Knock it off', meaning that he wanted out, and even though the practice display was far from over, the passenger's welfare was paramount and the normal proceedings were brought to a premature end. I began to wonder what I'd let myself in for.

During the same afternoon, two celebrity VIPs arrived by private plane from Burbank. One was Tim Berry, the producer of the TV show *Cheers*, while the other was Michael Dorn, a famous actor from *Star Trek*. Both had been invited to fly demonstration profiles in aircraft No 7 with Dave. Quite a large number of VIPs are offered this facility each year, as the 'Blue Angels' recognise the benefits of promoting their activities through what they call 'selected role models' in the sport, music, TV, radio and movie industry. As the public affairs staff confirm, it is an excellent way to get the 'Blue Angels' message to the widest possible audience: 'The familiar faces are a link through which the public can associate the professionalism and high standard of excellence the team embodies.' Likewise, many media representatives are also given a once-in-a-lifetime opportunity to fly in a 'Blue Angels' Hornet. As Capt. Ken Switzer ('Blue Angel' No 6) says: 'Flying media representatives is another way we can get our story out to the public. A television newscaster or radio or journalist celebrity has a unique perspective, and can communicate to a wide audience.' In his office, the public affairs officer, Lt Cdr Chuck Franklin, explained how he had dealt with countless visitors, and confirmed my feeling that Tim Berry was looking pretty nervous as he walked out to the F/A18 to take his flight. The vast majority of the VIPs have little or no flying experience and the Hornet is of course a pretty fearsome aeroplane to begin flying in. However, Chuck also assured me that just about everyone ultimately enjoys the experience and inevitably provided the team (and therefore the United States Navy and Marine Corps) with some excellent positive publicity. Tim Berry later returned, more than a little impressed with the capabilities of the Hornet, and on his way back to Burbank he was already considering ways in which he could incorporate the 'Blue Angels' in a future episode of *Cheers*. Michael Dorn then took to the air, noticeably less nervous, no doubt due to his experience as a private pilot. He returned forty-five minutes later, similarly awestruck by the F/A18B and, although he insisted that he'd had a great time, he commented that he suffered two G-LOC blackouts during the ride. His personal account would eventually make its way into the nation's media, capturing the interest of 'Trekkies' and more general television fans alike.

The next morning begins with the team's first display directly over the airfield at El Centro, and the four-ship again gets smartly airborne, right on time. After completing their manoeuvres, the two solos make their debut. With this performance complete, it's time for me to find my blue overalls while the team debrief. At 10.30 a.m., I join Cdr Wooldridge, Lt Larry Packer (No 2), Lt Doug Thompson (No 3), Lt Pat Rainey (No 4), Lt John Foley (No 5 lead solo), and Capt. Ken Switzer, USMC (No 6 opposing solo) in the briefing room. Although the team has already flown over 100 display practices during the past few weeks, the briefing is still long and detailed. Cdr Wooldridge runs through the entire sequence, step by step, tracing the formation's flight path over an aerial photograph of El Centro. Each team member offers his own input and, once everyone is completely happy, the debrief is finally over. Cdr Wooldridge ends with an upbeat tone: 'Let's make this a real good one.' He also reminds the team that this will be the first time during 1992 that the entire four-ship and solo display will be combined, and that despite my presence in the rear of No 7, the show should be flown exactly SOP (Standard Operating Procedure). Out on the flight line, aircraft No 7 has been towed into place, between aircraft Nos 3 and 6, having been substituted for the usual F/A18A in order to accommodate me. With all six aircraft parked in perfect formation, I climb aboard once more, a few minutes ahead of the rest of the team, so that the traditional pre-start walk-out can be completed properly without my interference. The team then march out to their aircraft in turn and perform a perfectly polished ceremony, even though today's audience is no more than a few dozen base personnel rather than the summer crowds of many thousands. Unlike many other display teams (such as the 'Red Arrows'), the 'Blue Angels' place great emphasis on the pre-take-off and post-take-off ceremony

too, involving both aircrew and ground crew. They consider themselves to be a Demonstration Squadron rather than an aerobatic team. The difference may appear to be subtle in principle, but the resulting show is consequently somewhat different to the typical European event, being designed to show the public as much as possible. The front seat of Hornet No 7 is soon occupied by Lt Pat Rainey and all six aircraft simultaneously start engines and taxi to the runway. As you might expect, even on the perimeter track the 'Blues' maintain a perfect (and very tight) formation, the positioning being just as perfect as that maintained in the air. Onto the runway, our four-ship nudges into position, powers up, and with another roar of engine power we roll forward in unison, with the two solos visible behind, waiting to begin their individual take-off runs behind us. With a speed of 135 knots on the MFD, all four aircraft lift gently into the air together and instantly the action begins with No 2 tucking-in on our starboard wing, and No 3 to port, as we race along just a few feet above the runway. The Boss's aircraft thunders directly overhead and I get an ultra-close look at the Hornet's main gear, which is still tucking away above and ahead of us. We immediately begin a gentle climb into the first loop and, from here on, my outside vision is almost continually filled with blue metal, the massive presence of a big, bulky Hornet just inches away to each side, and above me. The team pilots are proud to say that wingtip separation is usually just 36in, often overlapping in the vertical plane, but this separation varies second by second and on (rare) occasions the aircraft have been known to make physical contact, thankfully without any major consequences. From my viewpoint in the slot position, I'm right in the middle of the huge gaggle (or is it a swarm?), each aircraft bobbing up, down, left and right, as the pilots make countless minor control corrections in order to maintain their perfect formation. From the ground, the formation looks as smooth as glass. From up here, it looks very different – like four very large, very heavy warplanes thundering around the sky in perilously close proximity to each other, just milliseconds from collision. It looks very dangerous because it is. It's also fascinating, of course.

During some brief moments away from the display centre, while the various formation changes are made, I get an occasional opportunity to reload my film camera, taking care to keep my arms and legs well away from the control column and throttle. Care also has to be taken not to drop the camera or film, for if I lost contact with any objects, we'd be forced to leave the formation (and somebody would have the thankless task of dismantling the cockpit if necessary, to find the lost items). Safety in such critical circumstances is paramount, and loose objects run the risk of disabling flying controls in a very critical environment. Having reloaded my camera, I also get a quick glimpse of the two solo aircraft, visible over my shoulder, performing part of their routine low over the airfield. As they complete their manoeuvres we're already heading back to display centre, maintaining the continual flowing display profile that is seen by the spectator. At some stages during the routine, Pat Rainey advises me to place my camera firmly on my lap, before entering particularly severe manoeuvres in which a camera could (and has) caused injury to the photographer. As we enter the vertical pitchout, I'm warned to anticipate the onset of three gruesome 7.5g turns and, while Pat throws the Hornet through the recovery and rejoin manoeuvres, I grunt, groan and grasp, defiantly holding onto my camera. My cockpit is being filmed on video, as a miniature camera is installed above the instrument panel, allowing the PAO staff to record VIP passengers for their subsequent amusement (or embarrassment). After the flight we find that the video recorder failed during the more severe manoeuvres, but as our crew chief later informs me, this happens quite frequently as a tape recorder head is obviously not designed to mate with magnetic tape under severe g conditions. The show is quickly completed and we break for landing downwind to perform a perfect and precise return to the flight line, where the engines are stopped and the crew assemble in front of their mounts, still in perfect ceremonial harmony. Once the show routine has been completed, I too am invited to climb out of aircraft No 7 and walk (or stagger) down the flight line. The outstanding skill and professionalism of the 'Blue Angels' is a delight to behold, and the US Navy Flight Demonstration Squadron is one of the very best display teams in the world. Whether the 'Blues' could be described as the very best probably depends upon national pride and many subjective attitudes, but there's no doubt that when it comes to precise and demanding jet flying, the 'Blue Angels' are a hard act to beat.

Many miles away back in the UK, the RAF celebrated Her Majesty The Queen's 80th birthday in style on 17 June 2006 with a forty-nine aircraft flypast over Buckingham Palace. I was invited to see the flypast from an unusual vantage point by joining two crews from No 100 Squadron, who were tasked with the provision of 'whip' aircraft for the event. Whip aircraft are required to fly with the assembled aircraft in formation flypasts, in order to assist the participating pilots with their formation-keeping. The 2006 flypast was divided into a series of nine elements, all separated by forty seconds, and while the pilots within each element created their own formation positioning, the whip pilot and observer were called in to look at the formation from all angles in order to establish that it looks tidy and symmetrical. Once everything looks good, the individual pilots then take visual cues from adjacent aircraft (or the lead aircraft) and use these cues to maintain formation once the whip aircraft has gone. That's the theory, and it does work pretty well.

After visiting RAF Leeming on 13 June to deal with paperwork, medicals, kitting out, safety drills and so on, I return one day later to join Flt Lt Lance Millward in Hawk XX329 as 'Windsor Whip Two' for

ABOVE: On their way to salute HM the Queen, a quartet of Tornado F3 interceptors close in to make a tight formation off the wings of an E-3D Sentry from RAF Waddington. With impeccable precision, the formation will duly overfly Buckingham Palace to within a second of their allotted time. (*Tim McLelland*)

ABOVE: Over the sunny fields of Lincolnshire, Tornado GR4s formate on a Nimrod MR2 during a rehearsal for The Queen's birthday flypast. In order to avoid 'showing' on days prior to the actual event, rehearsal flights are directed towards Waddington, acting as a substitute Buckingham Palace for the preparations. (*Tim McLelland*)

the flypast's rehearsal. Departing from Leeming, we follow 'Windsor Whip One' (flown by Flt Lt James Harris with renowned photographer Geoff Lee in the back seat) down to the Norfolk coast, where we intercept the various flypast elements as they assemble in a long racetrack pattern centred on Southwold, on the Suffolk coast. Sadly, the weather conditions are poor; Leeming is under a blanket of thick cloud with rain, while the flypast route is sandwiched under varying thicknesses of cloud and haze. For the rehearsal, Waddington has been selected to represent Buckingham Palace, but as we head out over the cold, grey North Sea, we don't know whether the final run-in to Waddington will even be possible, so poor is the weather. However, the planned rehearsal continues, and with some careful radar and radio work we're guided towards the rear elements of the flypast, having decided to split our efforts into two halves; while 'Whip One' tackles the lead of the procession, we head for 'Vulcan Combine' – an E3D which is quickly joined by four Tornado F3s. Once they're all comfortably assembled into their planned formation, we're called in to take a closer look, and with a bit of feedback from us ('What do you think, Tim … number two needs to be forward and in slightly … number four is a little high …') the job's done pretty swiftly, and once completed we zip off to find the next formation, this time 'Tartan Combine' comprising a VC10 flanked by four Jaguars. Next on our list is 'Batman' formation – a gaggle of nine Tornado GR4s from Marham, which already look crisp and tidy even before we make any comment. It transpires that one of their airborne spare aircraft has already done a little improvised whipping before our arrival, so Flt Lt Millward merely congratulates the formation leader on their efforts (probably a wise decision considering that Marham's Station Commander is leading the formation) and we move on to 'Nimrod Combine' (a Nimrod flanked by four Tornado GR4s) while 'Whip One' goes on along the projected flypast route to take a look at the weather, having already whipped 'Typhoon Combine' (four Typhoons) and 'Fagin Combine' (two Typhoons and two Jaguars, minus the lead TriStar declared unserviceable before take-off). By this time, Waddington has been declared 'go' for the flypast, and we're reaching the end of our fuel reserves, so we leave the flypast route at Spalding and begin our return leg to Leeming, while the flypast elements continue to Waddington

before dispersing over Scampton. Unfortunately, 'Whip One' doesn't have quite enough fuel to make the journey, and pops in to RAF Coningsby to refuel.

With the rehearsal declared a success (the reserve days on Thursday and Friday not being required), our next task is to fly down to RAF Marham on Friday afternoon, so that we'll have sufficient fuel to stay with the formations along more of the Saturday flypast route that still centres on Southwold, but this time continues into Essex and finally Greater London. Saturday's weather is much better and after some problems removing our Hawk from one of Marham's temporary shelters (on which the door jammed partly closed) we make our way

LEFT: A sight designed to thrill: Nine Hawks from the legendary 'Red Arrows' team roar skywards in formation at RAF Waddington to begin another magnificent display. Once airborne, the three groups of aircraft will come together to create their familiar 'Big Nine' formation prior to commencing their performance. (*Tim McLelland*)

looking at the lead elements, comprising the Typhoons, the TriStar combine and the Nimrod formation. Because of their much slower operating speeds, the Battle of Britain Memorial Flight usually slot into the London flypast route from the Fairlop area, so they're not part of the whipping/assembling process. Likewise the rear element (this time the 'Red Arrows' and a Canberra) is treated as a separate package, slotting in from the South East. Consequently, we never get to see the much admired 'Reds' and Canberra formation, as we won't have enough fuel to go and find them, even if we had the time.

With whipping complete, and the formations already making their way along the route to London, our task is complete and we even have a couple of minutes left to play with before our fuel call. I decide that we should take a closer look at the solo C17 as I figure it will be the only chance of ever getting up close to one in the air. We make a quick rendezvous, tuck in, pull up and roll over the top to the left then back to the right, and that's it; time to head home with a burst of power and a sporty climb to 28,000ft to get over the airways. Boulmer Radar vectors 'Whip Two' back up at altitude to meet us and we make our way back to Leeming, listening in on the radio to the flypast aircraft as they complete their route and pass directly over Buckingham Palace.

The Hawk pilots have a primary task (whipping and weather checks) and they cannot devote any time to tasks like photography until their work is done. Unfortunately, by the time that work is done, the Hawk's fuel reserves are almost gone, so there's no time to wander around trying to get pictures. Consequently, you have to try and literally grab shots when aircraft come into range, even if it's only for a second or so, but even this can be a real challenge when you're clutching a camera with one hand and a reference chart with the other, trying to discuss the formation positions with the pilot, clicking the RT switch on and off whilst manoeuvring around the various elements. The lighting conditions change rapidly, so that one

to the runway, ready to depart behind the 'Batman' formation. Baking in the morning sun, we sweat out a few minutes while the Tornados lumber off in pairs, each making a full airfield circuit while another pair sets off to join the growing formation. Once the Tornados are all airborne we take the runway and head out to a much brighter North Sea, only to find that the rendezvous heights (around 3,000ft) are stuck in soup-like haze, which makes the task of picking up the various aircraft even harder than on the rehearsal day. Once again, my acclaimed long-range spotting skills (well, the pilots are impressed!) come in handy and we're soon with 'Typhoon formation' in the holding pattern. Having swapped jets to fly with Flt Lt Harris, I'm now minute you can be in gloomy haze and murk, and then a second later you're in brilliant sunshine, so you also need to keep an eye on your camera's settings. Likewise, there's also a need to keep a careful eye out for stray aircraft, disorientation, radio chatter and so on. There's a lot to keep you busy, and photography has to come pretty low on the list of priorities; in fact I probably got no more than five minutes of dedicated photography time in four hours of flying. Worse still, by the time the precious photography minutes come around, the elements are on their way to the 'target' (Waddington or the Palace in this case) and by then the formations are widely spaced out so that the pilots can relax until the final run-in commences, so you have to go with the circumstances you're presented with and get what you can. It's a sweaty, neck-twisting, stomach-clenching business and it can be very frustrating, but of course it's fascinating and just about as much fun as you can get on a Saturday afternoon. It's astonishing to see how much preparation, planning and hard work goes into creating just over five minutes of entertainment for Her Majesty, but judging by the television pictures, she was suitably amused!

It's not surprising that opportunities to participate in display performances are very rare. The RAF doesn't allow civilians to fly in any of its aircraft except in the rarest of circumstances, and permission to fly in public displays is an even rarer occurrence, not least because the aircraft crews have more than enough tasks on which to concentrate, without the additional concern of a passenger. However, I was lucky to participate in a few public events, such as the one described previously. Undoubtedly, the most thrilling were the flights that I made with the world-famous RAF Aerobatic Team, the 'Red Arrows'. There is something unique about the 'Reds', and despite being defined as merely a component squadron of the Central Flying School, they are of course a very special institution. Their base at RAF Scampton is perhaps rather less impressive than one might imagine. Even though Scampton is one of the RAF's most historic bases, it's only a shadow of its past. A great deal of the site has been sold off, and an ugly fence separates the remains of the RAF base from what is now a rather uninspiring housing estate. Many buildings have been demolished, but the huge hangars that once housed the Lancasters of the legendary Dambusters,

and then the mighty Vulcans, are still there. The 'Red Arrows' occupy one of these hangars, together with an administration block attached to it. Inside their offices, there's little to distinguish the surroundings from any other RAF squadron building. The crews and admin staff go about their business in their offices and briefing rooms, and the coffee machine continues to churn out its supplies inside a comfortable if rather unremarkable lounge. It's outside on the concrete apron that things are slightly different. No dull, grubby, camouflaged combat aircraft here. Instead, there's a line of gleaming, Signal Red-painted Hawks, all perfectly positioned and polished to perfection.

Preparing for a ride with the 'Red Arrows' is like preparing for a flight in any Hawk. The same flying clothing is needed, although the typical green-painted flying helmet is exchanged for a much brighter white example, marked with red trim. The pre-flight briefing is short, simply because there isn't much to discuss. The pilots know what they are going to do as they've done it countless times before, and the only variations that need to be discussed are the characteristics of different display sites, together with other issues such as airspace restrictions and weather conditions. The actual display, however, remains unchanged. It is built up during many weeks of winter training and modified into three distinct variations that are designed to cater for different weather conditions. For clear, sunny days, the team flies a full display, but if the weather and visibility are very poor, a flat display is substituted, keeping the team below the cloud cover. Somewhere between these two formats is the rolling display, and although the three display formats are clearly defined, the team leader can change between each format as each display progresses, should weather conditions change.

My first flight with the team is in Red Ten, the Hawk that's flown by the team manager as a 'spare' airframe. Our aim is to fly next to the assembled formation in order to take photographs, and although the experience is in effect a typical Hawk flight, it's undoubtedly strange and thrilling to close in on the nine Hawks as they sweep majestically over Lincolnshire. We then have to separate ourselves from the team as they commence their display routine, but from a comfortable height of 15,000ft above Scampton, we circle the airfield in a series of lazy turns while the team perform below us, barely perceptible against the landscape until their familiar trails of smoke billow into view. But this breathtaking view of the team is only a taster of the real excitement that comes from joining the team inside their nine-ship formation.

In order to see the display from every angle, I took a passenger seat in the lead aircraft flown by the Boss and then sampled two more displays from the vantage point of other aircraft within the team. Naturally, each display begins quite routinely, with a series of formation take-offs after which the groups of aircraft gather together into the familiar nine-aircraft combine prior to commencing each display performance. It's only then that you realise just where you are. With eight brightly painted Hawks beside you, there's no doubting that you're with what is perhaps the world's most famous display team and it's a thrill of almost unimaginable proportions to rush in formation over the display datum and hear the leader call 'smoke on' to begin a display performance. Having watched the 'Reds' from the ground more times than I can remember, it's quite extraordinary to begin a display from the opposite perspective, and even though it's difficult to see too much of what's going on down on the ground, it's easy to imagine that, as usual, an awful lot of people are gasping with excitement as the commentator announces '… the Royal Air Force Aerobatic Team, the Red Arrows … ' The performance starts with a steady climb into a wide loop, and from the lead aircraft the view is magnificent, with Hawks to either side, bobbing and weaving gently just a few feet away as we pull up into the blue sky and the lush fields below us withdraw into the distance. In the Hawk's rear-view mirror the long trail of white smoke can be seen, and in a matter of seconds we effortlessly begin a steady descent before levelling out and turning before the assembled spectators. From the lead aircraft nearly every manoeuvre is completed with almost effortless smoothness, and deliberately so, in order that the rest of the team pilots can keep their positions and follow the leader through each manoeuvre. There's a constant commentary from the leader, talking the team through every move, calling each formation change, every smoke-on or smoke-off call, and gentle, almost soporific comments as each turn, roll and loop is commenced.

The experience is of course both fascinating and joyful. But it isn't always like that, as I found out when I joined another display on a dull, cloudy and very windy day. Having opted to take a position out on the edge of the formation, the view from the cockpit is certainly different, with all eight aircraft out over my shoulder. Although the usual professional performance unfolds, it's almost impossible to imagine that the spectators below are seeing the same, precise formations that I've seen so many times before. From a team member's perspective, the view is very different. There are eight large Hawk jets, not in formation but seemingly in one great gaggle, all frantically writhing around in what looks like a completely disorganised mob. There's no perception of a formation here; indeed every aircraft seems to be moving independently of the others, up and down, left and right, back and forth, often looking as if the whole bunch of aircraft are about to crash together in one great crumpled heap. But of course my pilot is working hard to ensure that the movement between each aircraft is within astonishingly small limits. Even though from our perspective we are moving around quite significantly, from the viewpoint of the spectators below, the team is fixed into its usual, precise formation when we sweep by together, trailing our coloured smoke. The weather is terrible so the full display routine is impossible, and even the rolling display is a struggle to maintain. Consequently, we're obliged to fly some tight turns before the crowds, switching into different formations for each pass. It looks easy from the ground, but this is no joyride. The air is turbulent and we're thrashing around, seemingly clinging to our team-mates as we thunder around at low level. The turns are tight and the gforce is almost constant, pushing and grinding everyone into their seats. The pilots are grunting and groaning under the strain, often issuing expletives that their worshipping public below would probably be astonished to hear. But nobody can blame them as they're under almost unimaginable pressure, straining to keep each aircraft in place whilst turning hard at great speed, just a few hundred feet above the soggy fields that rush by us below. The formation turns are then replaced by even more violent manoeuvres as we begin cross-overs, roll-backs, bomb bursts and more, every move being made with crisp precision that demands skill and nerve. It requires a passenger to simply hold tight and wait for it all to stop. Every display is a punishing exercise that leaves any passenger exhausted, but the team members treat their task as a matter of routine. After landing, every display is debriefed in minute detail, every manoeuvre and every formation scrutinised on a large TV screen, and it isn't unusual for the team to be shaking hands, posing for photographs and signing autographs just minutes after they climb out of their cockpits. Any mere mortal would simply want to retire to a darkened room for a rest.

After one particularly punishing display routine I discussed the display with my pilot as we taxied in after landing. For an observer, the most surprising aspect of the 'Red Arrows' display is just how different it looks from inside the actual formation. It's almost impossible to describe how a nine-aircraft formation of Hawks in a tight, precise formation can look so very different when seen up close. I remarked that perhaps the 'Red Arrows' are a victim of their own success and skill. They do make their displays look incredibly simple and easy, and it's difficult to convey to any observer just how brutal and remarkably challenging it really is. The only way to demonstrate just what an astonishing task the 'Reds' perform would be to take every spectator and stick them in one of the Hawks so they could see for themselves, but obviously this cannot happen. The 'Red Arrows' pilots obviously take pride in the way in which the public are slightly fooled into thinking that their job is almost easy, but it's undoubtedly a shame that the thousands of spectators who see the 'Red Arrows' every year cannot understand just how incredibly difficult the pilot's job really is. It's undoubtedly the hardest and most punishing flying job that anyone could have, and the very fact that most people have absolutely no idea how difficult their task is, must be a testament to their skill and professionalism. It's certainly true that of all the flights I've made – so many that I've lost count – those with the magnificent 'Red Arrows' have been the most memorable.

RIGHT: This magnificent example of the Chipmunk is still flying, in the caring hands of a civilian owner. It is painted to represent an aircraft from the 'Sparrowhawks' aerobatic team, complete with the badge of the Central Flying School on the fuselage. (*Adrian Pingstone*)

RIGHT: Although no longer in regular military service, Chipmunk WK608 still flies with the FAA's Historical Flight at Yeovilton, acting as a trainer for pilots requiring 'tail wheel' experience, and as a communications aircraft. It retains standard FAA markings. (*Royal Navy*)

ABOVE: The Jet Provost entered RAF service in 1957, although acceptance trials began in 1955. The Central Flying School quickly formed an aerobatic display team with four T1 aircraft in 1957, operating as the 'Sparrows' throughout 1958. (*David Whitworth*)

LEFT: A fascinating pilot's-eye view of the RAF's 'Gemini Pair' Jet Provost team in action, the lead aircraft hanging inverted just feet above the second aircraft, with its pilot hard at work maintaining perfect formation. (*Tim McLelland collection*)

ABOVE AND RIGHT: Sequence photos showing a Tucano lowering its landing gear in flight. As can be seen, the main gear extends well ahead of the nose wheel leg. (*Tim McLelland*)

CLOCKWISE FROM LEFT: In recognition of part of the RAF's history, some Tucanos have received their own individual names. Moshoeshoe was King of Basutoland, and the Tucano's parent squadron is No 72 (Basutoland) Squadron, this unit having provided air defence of that region (now Lesotho) during both world wars. (*Tim McLelland*); High above the clouds, ZF142 illustrates the Tucano's all-black paint scheme complete with yellow trim. After a series of trials, this was found to be the best paint scheme for creating high conspicuity so that these training aircraft can be seen easily by other pilots sharing the skies as they go about their business. (*Tim McLelland*); A typical daily scene over Yorkshire as a Tucano races over the countryside on a low-level training exercise. The black-and-white propeller not only provides for safety on the ground but also adds conspicuity whilst in the air. (*Tim McLelland*)

LEFT: A sequence of images showing a Hawk from No 100 Squadron during its take-off roll at RAF Leeming. The landing gear can be seen retracting into its housings, and on the final frame Leeming's control tower is visible in the distance, alongside the A1(M) road. (*Tim McLelland*)

ABOVE: As the RAF's advanced trainer, the Hawk appears on the air show circuit almost every year, and the designated display aircraft are traditionally painted in dazzling show colours. For the 2010 season, two aircraft appeared in truly eye-catching (and patriotic) colours. (*Royal Air Force*)

RIGHT: Harrier T Mk 8 ZB603 on the flight line at Yeovilton. The Fleet Air Arm ended Harrier training during 2006, at which stage the Harrier T8 was retired. As can be seen, the aircraft's Sidewinder launch rails and external fuel tanks are painted grey, indicating that they have been 'borrowed' from a Sea Harrier FA2. (*Tom Cheney*)

RIGHT: A stunning image of two AV-8B Harrier II Plus aircraft, illustrating the revised nose contours, enabling the aircraft to accommodate an APG-65 radar, giving the Harrier the capability to operate beyond-visual-range missiles such as the AIM-120 AMRAAM. (*US Navy*)

ABOVE: A pair of Harrier II Plus aircraft pictured on board a USMC carrier. Naturally, the Harrier's capabilities made the aircraft ideal for carrier operations and perfectly suited to the USMC's requirements for a capable support aircraft. (*US Navy*)

LEFT: A very unusual view of an AV-8B Harrier as a battery of flares is launched. These flares were a standard fit on USMC Harriers, giving the aircraft a defensive capability against heat-seeking missiles. (*US Navy*)

LEFT: The RAF's association with the legendary Harrier ended in December 2010 when the last operational aircraft were withdrawn from service. The fleet was soon dismantled and transported to the Arizona desert, where it now acts as a spares source for the USMC. (*Gareth Brown*)

RIGHT: XW529, assigned to test and research duties, retained its factory-applied paint scheme throughout its operational life. It was assigned to various establishments including Boscombe Down, Hatfield and Scampton, and remained in use until 1988. (*Tim McLelland collection*)

LEFT: A quartet of Buccaneers over the Moray Firth, just days before the type was retired from RAF service. Two of the aircraft are painted in a light grey camouflage scheme which was in the process of being introduced on the fleet at the time of the aircraft's withdrawal. (*Tim McLelland*)

LEFT: An intimate view of the Buccaneer in its landing configuration over a typical maritime environment in which the aircraft was designed to operate. (*Tim McLelland*)

RIGHT: Typical fast-jet (including Buccaneer) head gear for RAF pilots. The standard helmet incorporates both tinted and clear hinged visors and a clip-on oxygen mask on which the radio communications switch can be seen. (*Royal Air Force*)

LEFT: A bleak scene in Lithuania as a C-17 prepares to depart for the United States, loaded with cargo. The C-17 provides the US Air Force with true global reach, enabling large loads of cargo and passengers to be deployed at will across the world. (*US Air Force*)

LEFT: Out in the Nevada desert a C-17 demonstrates its remarkable short field landing capability, which enables the aircraft to operate from small and unprepared strips with ease thanks to its innovative design, incorporating high lift devices and thrust reversal systems. (*US Air Force*)

RIGHT: Inside an RAF C-17's capacious cargo hold a Merlin helicopter is prepared for transportation. Given the size and weight of the huge Merlin, this is a graphic illustration of the C-17's lift capability. (*Royal Air Force*)

LEFT: This unloaded C-17 interior illustrates the sheer size of available internal space and the flexible nature of the aircraft's capabilities; fold-out seats for troops are fitted along the fuselage sides. The flight deck is visible at the forward end of the hold. (*Jeff Eddy*)

ABOVE: Low over Loch Ness, a WSO's-eye view of a Tornado mission, as three aircraft race south on a training mission. In addition to 1,000lb bombs, the Tornado's internal cannon is clearly visible under the cockpit, below the bolt-on refuelling probe fairing. (*Tim McLelland*)

ABOVE: Jaguars from No 6 Squadron over the North Sea during a training mission, passing Hartlepool en route to their home base at RAF Coningsby in Lincolnshire. The gloomy weather conditions are typical of the kind of environment in which the Jaguar often operated. (*Tim McLelland*)

LEFT: This unusual view of two Jaguars (taken from another Jaguar rolling above them) illustrates the type's unique arrangement for self-defence missiles. The launch rails (normally fitted on underwing hard points on combat aircraft) are attached to the wing's upper surfaces, enabling the lower hard points to be available for additional stores, most notably external fuel tanks, as carried by these aircraft. (*Tim McLelland*)

RIGHT: The view from the rear cockpit of a Jaguar T4, looking through the aircraft's HUD, illustrating a speed of 474 knots and a height of just 285ft. A second Jaguar is just visible above the pilot's ejection seat. (*Tim McLelland*)

RIGHT: Lightning T Mk 5 XS422 was operated by the Empire Test Pilots' School at Boscombe Down and was for many years the pride of their aircraft fleet. Now in civilian hands, it is being restored by a group of enthusiasts in the USA, where it is hoped it will fly again in due course. (*Tim McLelland collection*)

RIGHT: Two Phantom FGR Mk 2 aircraft from No 56 Squadron racing across the Cumbrian hills, trailing moisture vortices from their wing tips. Sidewinder missiles are visible on the aircraft's underwing weapons stations. (*Tim McLelland collection*)

LEFT: Phantom FGR Mk 2 XT903 was one of a few RAF Phantoms capable of being reconfigured as a dual-control aircraft if required for training purposes. Although the Phantom's forward cockpit controls were not completely replicated in the rear cockpit, flying controls could be installed so that an instructor could operate the aircraft, albeit with some difficulty. In practice, 'twin-stick' Phantoms were only rarely operated. (*Tim McLelland collection*)

ABOVE: Fully inverted, high over the Californian desert, a 'Blue Angels' Hornet displays the team's traditional blue colours and yellow trim, a design that has been applied to a series of aircraft that the team has operated over many years, ranging from Bearcats through to Tigers, Phantoms and Skyhawks. (*US Navy*)

LEFT: This is a typical view from the pilot's seat in an F/A-18 Hornet during a 'Blue Angels' display. The proximity of the aircraft to each other is remarkable and startling for an unprepared observer, with just inches separating the wingtips of each aircraft in the formation. Not surprisingly, aircraft do occasionally make contact with each other, although mercifully without any catastrophic results. (*US Navy*)

RIGHT: This view from the rear cockpit of a 'Red Arrows' Hawk illustrates the Miniature Detonating Cord (MDC) fixed in the Hawk's canopy, designed to shatter it if the ejection seat is fired. Although prominent, it is not obtrusive to the pilot's view, as the formation streaks towards RAF Akrotiri, Cyprus, to begin a display rehearsal. (*Royal Air Force*)

RIGHT: Over the top of a loop, the view from the Hawk's rear cockpit is excellent, with a clear view out to left and right, and forwards above the front cockpit ejection seat. Only the MDC partially blocks the upward view, and the large canopy frame holds two very useful rear view mirrors. (*Royal Air Force*)

RIGHT: Hawk XX266, better known as 'Red Nine', over the sands of the Dubai Desert during a visit to the Middle East. The 'Royal Air Force' corporate-style titling is a relatively recent and modern addition to the team's aircraft, but thankfully the more modest badge of the Central Flying School remains on the aircraft's forward fuselage. (*Royal Air Force*)

LEFT: A magnificent late afternoon image of the 'Red Arrows' heading north from their base at RAF Scampton in Lincolnshire for a display in Prestwick, many miles away in Scotland. With white smoke trailing for the cameraman, the specially designed smoke ejectors are visible above the aircraft's jet exhaust. (*Royal Air Force*)

INDEX

If you enjoyed this book, you may also be interested in …

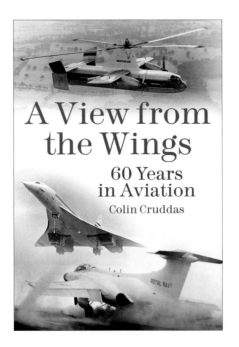

A View from the Wings

COLIN CRUDDAS

A View from the Wings is unique, recalling a wartime boyhood in which aircraft flying constantly overhead played a large part. This experience led to a lifetime career in the aviation industry both in the UK and overseas, such as the US and South Africa. Mixed with events of a more personal nature, often coated with whimsical humour, the author has evocatively captured the rise and demise of Britain's aircraft industry in the post-war period. In setting out to be non-technical, *A View from the Wings* will appeal to those whose memories embrace the sound barrier-breaking years and the leap of faith and technology that saw Concorde defeat the Americans in the race to produce a practical supersonic airliner.

978 0 7524 7748 0

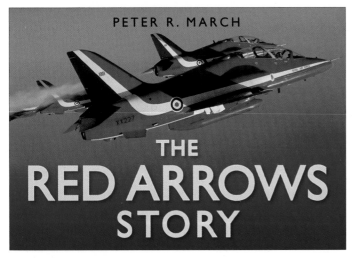

The Red Arrows Story

PETER R. MARCH

Acclaimed as the RAF's most famous formation aerobatic display team, the 'Reds' – as they are affectionately known – have attracted a truly international following. Here is the story of the Red Arrows, from their tentative beginnings as the Yellowjacks flying Folland Gnats in 1964, through the formation of the Red Arrows in 1965 to their conversion to the scarlet-painted BAE Hawks that they fly today.

978 0 7509 4446 5

Visit our website and discover thousands of other History Press books.

www.thehistorypress.co.uk